# World Sustainability Series

**Series Editor**

Walter Leal Filho, European School of Sustainability Science
and Research, Research and Transfer Centre "Sustainable Development
and Climate Change Management", Hamburg University of Applied Sciences,
Hamburg, Germany

Due to its scope and nature, sustainable development is a matter which is very interdisciplinary, and draws from knowledge and inputs from the social sciences and environmental sciences on the one hand, but also from physical sciences and arts on the other. As such, there is a perceived need to foster integrative approaches, whereby the combination of inputs from various fields may contribute to a better understanding of what sustainability is, and means to people. But despite the need for and the relevance of integrative approaches towards sustainable development, there is a paucity of literature which address matters related to sustainability in an integrated way.

**Notes on the quality assurance and peer review of this publication**

Prior to publication, the works published in this book are initially assessed and reviewed by an in-house editor. If suitable for publication, manuscripts are sent for further review, which includes a combined effort by the editorial board and appointed subject experts, who provide independent peer-review. The feedback obtained in this way was communicated to authors, and with manuscripts checked upon return before finally accepted. The peer-reviewed nature of the books in the "World Sustainability Series" means that contributions to them have, over many years, been officially accepted for tenure and promotion purposes.

Ayyoob Sharifi · John Lee Candelaria ·
Dahlia Simangan · Shinji Kaneko
Editors

# Navigating Peace and Sustainability in an Increasingly Complex World

*Editors*
Ayyoob Sharifi ⓘ
The IDEC Institute
Hiroshima University
Higashihiroshima, Hiroshima, Japan

Dahlia Simangan
The IDEC Institute
Hiroshima University
Higashihiroshima, Hiroshima, Japan

John Lee Candelaria
Graduate School of Humanities and Social
Sciences
Hiroshima University
Higashihiroshima, Hiroshima, Japan

Shinji Kaneko
The IDEC Institute
Hiroshima University
Higashihiroshima, Hiroshima, Japan

ISSN 2199-7373          ISSN 2199-7381  (electronic)
World Sustainability Series
ISBN 978-981-97-8771-5       ISBN 978-981-97-8772-2  (eBook)
https://doi.org/10.1007/978-981-97-8772-2

# Preface

The interconnections between peace and sustainability are widely acknowledged in academic and policy spheres, a recognition reaffirmed by the adoption of the 2030 Agenda for Sustainable Development. However, many initiatives addressing peace and sustainability continue to operate independently, overlooking their mutually beneficial relationship. In an increasingly complex world, it is essential to integrate peace-building efforts with sustainable development strategies to effectively address the multifaceted challenges associated with global transformations at environmental, social, political, and economic levels.

*Navigating Peace and Sustainability in an Increasingly Complex World* aims to explore different avenues that bridge the promotion of peace and sustainable development in an integrative manner. This volume builds upon two previous works, *Integrated Approaches to Peace and Sustainability* and *Bridging Peace and Sustainability Amidst Global Transformations*, published in early and late 2023, respectively. These earlier volumes successfully drew global attention to the relationship between peace and sustainability, paving the way for more in-depth exploration.

This book features a collection of papers presented at the NERPS 2024 conference, organized by the Network for Education and Research on Peace and Sustainability (NERPS) (Fig. 1). The conference took place from March 6 to 9, 2024, at Hiroshima University, Japan, with the primary aim of fostering transdisciplinary research and policy discussions on the intersection of peace and sustainability. By providing a platform for dialogue and collaboration among diverse stakeholders, the conference promoted further research and initiatives exploring the interconnectedness of these crucial global issues. With participation of around 175 individuals, including students, researchers, and practitioners across various fields and sectors, a wide range of topics related to the peace-sustainability nexus was explored.

**Fig. 1** The opening ceremony of NERPS 2024

*Navigating Peace and Sustainability in an Increasingly Complex World* explores the intricate relationship between sustainability and peace, offering both theoretical analysis and real-world examples. It recognizes that these concepts are multifaceted and dynamic, constantly evolving to address new challenges in our increasingly complex global landscape. The book examines how sustainability and peace intersect within different social, cultural, economic, and political contexts, elucidating the various factors that contribute to the realization of sustainable and peaceful societies in the face of growing global complexities.

We anticipate that this edited volume will provide valuable insights into the intricate relationship between peace and sustainability, particularly in the context of our increasingly complex world. Nevertheless, there remains a pressing need for further exploration to fully comprehend the interplay between these two concepts as global challenges continue to evolve. We encourage researchers and policymakers to participate in upcoming conferences and events organized by NERPS, where discussions on the peace-sustainability nexus can be expanded upon and adapted to address emerging global complexities. As we navigate the challenges of our increasingly interconnected and complex world, the integration of peace and sustainability becomes not just desirable but essential.

Higashihiroshima, Japan

Ayyoob Sharifi
John Lee Candelaria
Dahlia Simangan
Shinji Kaneko

# Contents

# About the Editors

**Ayyoob Sharifi** is Professor at the IDEC Institute, Hiroshima University. He also is a core member of the Network for Education and Research on Peace and Sustainability (NERPS). His research is mainly focused on urban climate change mitigation and adaptation. Ayyoob actively contributes to global change research programs such as the Future Earth and has served as a lead author for the Sixth Assessment Report (AR6) of the Intergovernmental Panel on Climate Change (IPCC). The ultimate goal of his education and research activities is to inform actions toward building sustainable and peaceful communities.

**John Lee Candelaria** is Assistant Professor at the Graduate School of Humanities and Social Sciences, Hiroshima University and a research fellow of the Network for Education and Research on Peace and Sustainability (NERPS). He completed his Ph.D. from the International Peace and Coexistence Program at Hiroshima University and has obtained both his bachelor's and master's degrees in History from the University of the Philippines Diliman. He researches topics in peace and conflict, including war propaganda, heritage and memory politics, peace processes, and the intersection of peace and sustainability, with a geographical emphasis on the Philippines and Southeast Asia.

**Dahlia Simangan** is Associate Professor of International Relations at Hiroshima University. Her research interests in peace and conflict include topics on post-conflict peacebuilding, the relationship between peace and sustainability, and international peace and security in the Anthropocene. She is the author of *International Peacebuilding and Local Involvement: A Liberal Renaissance?* (Routledge, 2019) and published her work in leading International Relations and Peace Studies journals. She is Associate Editor of the *Review of International Studies*, Assistant Editor of *Peacebuilding*, and a member of the Planet Politics Institute.

**Shinji Kaneko** is the Executive Vice President for Global Initiatives and a Professor at the Graduate School of Humanities and Social Science at Hiroshima University. He graduated from the School of Engineering at Kyushu University majoring in water engineering (Dr. of Engineering). Immediately after the completion of his doctoral program, he joined the Institute for Global Environmental Strategies (IGES) in 1999 as a researcher and conducted research on urban climate policy in Asian megacities for three years. He also worked at global Change System for Analysis, Research and Training (START) as a Research Fellow for three years from 2005. In 2018, he was appointed as the Director of Network for Education and Research on Peace and Sustainability (NERPS) at Hiroshima University. He has conducted numerous policy research on natural resources, energy, and the environment in developing countries.

# Navigating Peace and Sustainability in an Increasingly Complex World

Ayyoob Sharifi⬩, John Lee Candelaria, Dahlia Simangan, and Shinji Kaneko

**Abstract** This introductory chapter explores the intricate relationship between peace and sustainability in an increasingly complex world. It highlights the escalating challenges of global conflicts, climate change, and socio-economic instabilities that threaten both peace and sustainable development. The chapter emphasizes the interconnectedness of these concepts, as recognized in the 2030 Agenda for Sustainable Development, particularly SDG 16. It discusses the evolution of peace and sustainability research, noting the historical focus on conflict resolution and resource management, respectively, while advocating for a more holistic approach. The chapter synthesizes insights from various contributions to the third conference of the Network for Education and Research on Peace and Sustainability (NERPS 2024), covering topics such as environmental ethics, decolonizing environmental peacebuilding, social entrepreneurship in post-conflict regions, energy transitions, and environmental diplomacy. These diverse perspectives illuminate the multifaceted nature of the peace-sustainability nexus. The chapter concludes by outlining key implications for research and practice, including the promotion of indigenous entrepreneurship, the need for equitable energy transition policies, and the importance of stakeholder engagement in decision-making processes. This overview chapter sets the stage for a deeper exploration of strategies to navigate peace and sustainability challenges in our complex global landscape.

**Keywords** Peace · Sustainability · Nexus · Complexity · Conflict · SDGs

A. Sharifi (✉) · D. Simangan · S. Kaneko
The IDEC Institute, Hiroshima University, Higashihiroshima, Japan
e-mail: sharifi@hiroshima-u.ac.jp

J. L. Candelaria
Graduate School of Humanities and Social Science, Hiroshima University, Higashihiroshima, Japan

# 1 Introduction

Since the end of the Cold War, violence has reached unprecedented levels, indicating that the global conflict landscape has become increasingly complex (PRIO 2024). The past three years saw more conflict-related deaths than any time in the last three decades. The ongoing Ukraine-Russia conflict has inflicted severe human suffering and disrupted global food and energy supplies, exacerbating economic instability (Bin-Nashwan et al. 2024). Concurrently, the enduring Israel-Palestine conflict has seen renewed violence and political discord, fueling polarization and further complicating the path to peace (Bailliet 2024). Tensions in the Asia–Pacific, particularly around the South China Sea and the increasing assertiveness of China, also pose significant threats to regional stability (Koga 2023). Diplomatic efforts, including attempts to revive the Iran nuclear deal following the expiration of United Nations restrictions on Iran's missile and drone programs in 2023, have produced mixed results, reflecting the complex and often intractable nature of modern geopolitical disputes. Furthermore, the continued development and modernization of nuclear arsenals by various nations (SIPRI 2023) heighten global security concerns, underscoring the urgent need for renewed international arms control efforts.

The intensifying climate crisis paints an equally grim picture. 2023 marked the warmest year on record (Poynting and Rivault 2024), with extreme weather events devastating communities worldwide (Berman 2023), highlighting the urgent need for robust climate action. Despite international meetings such as COP28 emphasizing enhanced commitments for greenhouse gas emission reduction, the gap between pledges and tangible actions remains significant (Arora 2024). Biodiversity loss continues unabated, driven by habitat destruction, pollution, and climate change, despite international frameworks like the Kunming-Montreal Global Biodiversity Framework of 2022 aimed at curbing these trends. Progress toward the Sustainable Development Goals (SDGs) has been uneven, with notable improvements in areas such as clean energy access but substantial challenges in reducing inequality and ensuring sustainable consumption (United Nations 2024b). The aftermath of the COVID-19 pandemic has further complicated these efforts. These realities point to the need for a more integrated approach that addresses health, economic resilience, and environmental policy in tandem.

These global events are critical issues of our time, and among these, peace and sustainability emerge as paramount concerns. While traditionally viewed as the absence of conflict, the concept of peace has evolved to encompass positive peace (Galtung 1969), which includes social justice, economic equity, and ecological balance (Galtung 1990). Similarly, sustainability involves environmental protection, economic development, and social equity through an approach that "meets the needs of the present without compromising the ability of future generations to meet their own needs" (WCED 1987). Although distinct in their focus, we find that peace and sustainability are deeply interrelated, each influencing and being influenced by the other (Sharifi et al. 2021a, b). Researchers, practitioners, and policymakers widely recognize their interdependence and mutual reinforcement. The 2030 Agenda for

Sustainable Development, unanimously adopted in 2015, marked a significant global commitment to prioritizing human well-being and ecological balance, a recognition that peace and sustainability are mutually interconnected. Central to this agenda is SDG 16, which focuses on promoting peaceful and inclusive societies, guaranteeing access to justice for everyone, and building well-functioning, transparent, and participatory institutions across all levels of governance. SDG16 underscores the critical role of peace and justice in achieving all other sustainable development goals. Without peaceful and just societies, progress in areas such as poverty reduction, health, education, and climate action becomes inherently fragile and unsustainable.

The SDGs provide a comprehensive framework for examining the nexus of peace and sustainability. However, the prevailing scholarship on peace has traditionally concentrated on the dynamics of war and violence, often at the expense of understanding the complex mechanisms that foster and sustain peace (Gledhill and Bright 2019). This limited scope stems from the historical emphasis on conflict resolution, prioritizing the cessation of hostilities and the establishment of formal agreements (Tidwell 2001). Consequently, the ongoing processes and structures crucial for durable peace, such as social cohesion, justice, equity, and inclusive institutions, are frequently marginalized in post-conflict settings (Stewart 2009). Furthermore, the inherent conceptual ambiguity of peace, often perceived as a normative ideal rather than an empirical phenomenon (Richmond 2008), contributes to its underrepresentation in scholarly discourse. As a result, grassroots peacebuilding initiatives, local conflict resolution practices, and the role of everyday interactions in maintaining social harmony are often overlooked.

Similarly, much of the current sustainability research gravitates towards the sustainable extraction of natural resources, emphasizing the management and conservation of physical assets like water, minerals, and forests (Rockström et al. 2009). While indispensable, this approach tends to overshadow the broader facets of sustainability, particularly those concerning human-ecosystem interactions and long-term adaptability. The focus on resource extraction often neglects the vital social and cultural dimensions of sustainability, including how communities adapt to environmental shifts, preserve traditional ecological knowledge, and ensure intergenerational equity (Berkes 2018). The conflation of the terms "sustainability" and "sustainable development" further muddies the conceptual waters (Redclift 2005) and makes their competing language challenging to differentiate. Purvis et al. (2019) differentiate the two concepts as follows: sustainability, with deeper historical roots and less baggage, broadly denotes the capacity to endure and maintain ecological balance over time and encompasses a broad spectrum of considerations that consider environmental integrity, economic viability, and social equity. On the other hand, sustainable development is more focused on achieving sustainability, as it historically emerged and gained prominence through international policy frameworks. This semantic overlap can lead to conceptual ambiguity and practical challenges in policy and discourse, obscuring their interconnected goals. Consequently, efforts to operationalize, assess, and measure sustainability can become fragmented and inconsistent (Candelaria et al. 2023; Sala et al. 2015), hindering the development of coherent policies and practices.

Nevertheless, the global community has attempted to address peace and sustainability concerns for decades. Since the 1972 United Nations Conference on the Human Environment, countries have set aside national interests to prioritize environmental concerns, highlighting the intrinsic relationship between peace and sustainability. This conference resulted in the Stockholm Declaration and the creation of the United Nations Environment Programme (UNEP), which were significant steps in international environmental cooperation. The Stockholm Declaration laid down a set of principles for the preservation and enhancement of the human environment, recognizing the necessity of sustainable development to achieve global peace. The UNEP, established as a direct outcome of the conference, is the world's leading environmental authority that charts the course for global environmental efforts and ensures that environmental considerations are consistently addressed across all UN sustainable development initiatives.

While these advancements were critical, they did not lead to immediate and substantial policy momentum until the next decade. Policy entrepreneurs began emphasizing the intricate links between environmental sustainability and global peace during this period. The Brundtland Commission's *Report on Environment and Development: Our Common Future* in 1987 was pivotal in this shift. The report expanded the notion of security to include environmental stress, underscoring that traditional security measures, which focused predominantly on political and military solutions to sovereign threats, were inadequate to address the pervasive nature of environmental insecurity. Environmental issues, after all, do not recognize national or regional borders, posing transboundary threats that necessitate collective action and comprehensive policies. The recognition of the environment-security-development nexus significantly influenced subsequent international policy frameworks. The 1992 Earth Summit in Rio de Janeiro was instrumental in integrating sustainable development into the global agenda. The summit produced key documents such as the Rio Declaration on Environment and Development, Agenda 21, and the establishment of the Commission on Sustainable Development. These initiatives aimed to balance economic growth with environmental protection and social equity, setting a precedent for future international meetings. Subsequent conferences have continued to build upon the foundations laid in earlier decades. The Millenium Development Goals (MDGs) adopted in 2000 were succeeded by 17 SDGs outlined in 2015, which identified "areas of critical importance for humanity and the planet" (UN 2015: 1), addressing a broad spectrum of issues from poverty eradication and health improvement to climate action and ecosystem preservation. The SDGs recognize that achieving sustainability requires a holistic approach that balances economic growth, social inclusion, and environmental protection. This integrated perspective underscores the importance of addressing interconnected global challenges through comprehensive, coordinated efforts.

Despite fostering significant policy dialogue, the comprehensive integration of the SDGs within political institutions and legislative frameworks remains underwhelming and remarkably challenging. The 2024 SDG Report indicates that nearly half of the 17 targets show minimal or moderate progress, while over one-third are either stalled or regressing (United Nations 2024b), with the UN Secretary-General António Guterres likening the report as the world "getting a failing grade" (UN 2024a). Various factors, such as the COVID-19 pandemic, climate change, and geopolitical conflicts, have severely impeded efforts to achieve these goals. The pandemic, particularly, has strained public health systems and economies worldwide, redirecting attention and resources away from sustainable development initiatives. Climate change continues to exacerbate environmental stress, leading to more frequent and severe natural disasters, which undermine development gains. Additionally, ongoing geopolitical conflicts divert critical resources and focus from long-term sustainability objectives. Regional conflicts, such as the Ukraine-Russia War and the Israeli-Palestinian conflict, pose significant challenges to peace and sustainability. The cyclical nature of violence, political stalemates, and resource scarcity undermine efforts towards lasting peace and sustainable development. Environmental degradation, water scarcity, and infrastructure destruction during conflicts exacerbate humanitarian crises, highlighting the intricate link between peace and sustainability.

In an era of unprecedented complexity, escalating crises threaten human society and ecological integrity. The urgency for collaborative solutions to foster peace and promote sustainability has never been more pronounced. Academic research plays a vital role in informing and shaping these solutions. To fully grasp the interconnectedness of peace and sustainability, it is crucial to adopt, on one hand, a transdisciplinary approach that integrates insights from multiple perspectives. On the other hand, a synergistic approach that promotes collaboration among researchers and policymakers is equally essential. These approaches have been at the core of Hiroshima University's Network for Education and Research on Peace and Sustainability and the driving force behind its third international conference in March 2024. Held at the Higashihiroshima campus, the conference aimed to reaffirm and invigorate the commitment to interdisciplinary, multidisciplinary, and transdisciplinary approaches in addressing the pressing global challenges to peace and sustainability. This volume features contributions from that conference, underscoring the enduring necessity and critical urgency for such collaborative research in navigating the complex issues of our time.

## 2 Efforts Toward Navigating Peace and Sustainability

This volume addresses multiple environmental, social, economic, and institutional dimensions of peace and sustainability with insights from various contexts. A synthesis of the chapters, depicting efforts toward navigating peace and sustainability is provided here.

Chapter 2 explores the significance of integrating Islamic environmental ethics into broader sustainability discussions. It emphasizes the responsibility of humans to protect and preserve the environment, drawing on ethical principles from the Qur'an and Hadith. Key concepts like Tawhid and Ihsan guide Muslims towards environmental stewardship, highlighting the interconnectedness of all living beings.

The chapter advocates for incorporating Islamic values into education and organizational practices for sustainable environmental management. It also discusses the impact of artificial intelligence on environmental ethics and the role of professions like sanitary engineering in promoting ethical behavior towards the environment. The methodology used includes a Systematic Literature Review to comprehensively study Islamic environmental ethics.

Chapter 3 discusses the importance of decolonizing environmental peacebuilding by integrating indigenous perspectives, focusing on the Sama-Bajau maritime communities in Southeast Asia. It highlights how colonial legacies, structural violence, statelessness, and cultural, epistemic, and ecological violence impact the Sama-Bajau. The chapter emphasizes the need to recognize indigenous knowledge, address power dynamics, and promote sustainable practices. Decolonizing environmental peacebuilding involves respecting diverse worldviews, centering Indigenous voices, and fostering inclusive, culturally sensitive approaches. By acknowledging the agency and resilience of marginalized communities like the Sama-Bajau, promoting relational ways of knowing, and embracing pluriversal peacebuilding, a more just and sustainable future can be envisioned. The chapter advocates for collaborative efforts, regional cooperation, and a shift towards holistic peacebuilding strategies that consider social, economic, and environmental dimensions for lasting positive peace.

Chapter 4 is focused on Aretes Style, a social enterprise in post-conflict Marawi City, emphasizing its role in promoting peace, empowerment, and economic development. Through collaborations with organizations like UNDP, UNICEF, and ILO, Aretes Style leverages technology and social media platforms to enhance its impact on marginalized communities, particularly women weavers and internally displaced individuals. The enterprise's inclusive approach, despite challenges in the socio-political landscape, underscores the importance of supportive institutional environments for social enterprises. Aretes Style's success is attributed to its ability to navigate the complex institutional landscape, align with cultural norms, and foster community acceptance, showcasing the transformative power of social entrepreneurship in driving sustainable development, peacebuilding, and empowerment in conflict-affected regions.

Chapter 5 covers key themes related to Vietnam's energy transition, emphasizing the importance of institutional frameworks, climate resilience, human resource development, science and technology improvement, and more. It discusses how governmental decisions address these themes with varying focus. Equity considerations like distributional justice, economic equity, and power dynamics are analyzed, highlighting the need for clearer targets and strategies. The chapter explores Vietnam's energy transition policies, focusing on equity dimensions and the nation's commitment to low-carbon energy sources. It discusses the importance of addressing socio-economic burdens during the transition and integrating environmental justice principles. The study identifies major decisions related to green energy transition, power development plans, and climate change strategies issued by various authorities. This comprehensive overview aids in understanding Vietnam's energy and environmental policies.

Chapter 6 explores the challenges faced by the apparel industry in Bangladesh during the COVID-19 pandemic in implementing social sustainability practices. These challenges range from meeting government and foreign buyers' requirements to issues like reduced workers' salaries, factory shutdowns, and competitive pressures. Motivators for integrating social sustainability practices include ensuring worker safety, enhancing company image and profitability, and addressing workers' psychological well-being. The study emphasizes the importance of addressing these challenges and motivators to enhance social sustainability practices in the apparel industry. It also discusses conflicts between livelihoods and health safety, regulatory compliance, resource management, and shifting consumer demands. The chapter aims to provide insights for policymakers and industry stakeholders to implement sustainable practices during crises, particularly in developing countries like Bangladesh.

Chapter 7 is focused on the Rohingya Refugee Crisis in Bangladesh, originating from the 2017 genocide in Myanmar. It highlights challenges faced by the Rohingya people, such as economic crises, conflicts, security concerns, and environmental impacts in Cox's Bazar. The chapter, based on a review of 14 articles, emphasizes issues affecting peace and sustainability for both the Rohingya and host communities. It stresses the necessity of international support for peaceful repatriation and livelihoods for the Rohingya, addressing the lack of focus on the host community's perspective and the limited resources in densely populated Bangladesh. The crisis has resulted in social insecurity, environmental damage, and economic losses, necessitating actions to mitigate these impacts. The document proposes solutions like prioritizing repatriation with Myanmar, addressing security concerns, combating illegal activities, restoring ecosystems, and fostering collaboration between Bangladesh, Myanmar, and international communities to ensure peace and sustainability in the region.

Chapter 8 explores the reintegration of Indigenous migrant workers in the Philippines through social entrepreneurship. It highlights the challenges faced by returning Indigenous migrants, the government's reintegration programs, and the role of social enterprises in promoting economic empowerment and cultural preservation. The Cordillera region, known for its intricate handwoven textiles, serves as a case study for weaving entrepreneurship. Social enterprises in Cordillera focus on empowering weavers economically, preserving cultural heritage, and fostering community development. These enterprises provide fair compensation, support skill development, and ensure timely payments to weavers, bypassing exploitative practices. By engaging in sectors like agriculture, tourism, and handicrafts, returning Indigenous migrants can contribute to local economies while preserving their cultural identity. The collaborative efforts of stakeholders, government support, and a focus on social good are essential for the success of Indigenous social enterprises. Overall, social entrepreneurship emerges as a promising avenue for promoting economic development, cultural sustainability, and community well-being among returning Indigenous migrant workers in the Philippines.

Chapter 9 extensively covers the evolution of transnational agrarian movements, particularly focusing on UNORKA-Mindanao and their role as development partners for small-scale banana farmers in Davao del Norte, Philippines. It discusses strategies for empowering farmers, challenges faced such as anti-reform sentiments and legal disputes, and the importance of grassroots movements in advocating for land rights and social justice. The narrative also explores conflicts and resistance in banana farming, highlighting disputes between farmer organizations and influential families/companies, such as Lapanday Foods Corporation and Marsman Estate Plantation, Inc. The chapter emphasizes the significance of peasant politics, legal advocacy, and economic initiatives in addressing land rights issues and promoting genuine agrarian reform in Davao del Norte. Overall, it underscores the complexities of farmer empowerment, the challenges in securing land rights, and the importance of systemic change for advancing social justice in the context of the global agribusiness industry.

Chapter 10 discusses Taiwan's evaluation in the Government Defense Integrity Index (GDI) 2020 report, highlighting its Band "B" rating overall and Band "A" rating for personnel corruption risk. It emphasizes the importance of using GDI results to enhance military integrity through strategies like protecting whistleblowers, ensuring transparent recruitment and promotion processes, and implementing anti-corruption policies. The document investigates the correlation between low corruption and factors like globalization and stable economies, proposing evidence-based strategies for improvement based on GDI evaluations and international comparisons. Taiwan excels in areas like public commitment to integrity but needs to address shortcomings in appointment systems and promotion processes. Strategies to reduce corruption risks among defense personnel include ethical leadership, training, legislation, and effective human resource management, with a focus on data-driven personnel selection and the crucial role of whistleblowers in exposing corruption. The ultimate goal of these strategies is to enhance Taiwan's defense integrity before GDI 2025.

Chapter 11 explores Morocco's agricultural sector, emphasizing its economic significance and challenges. The country's diverse geography and agricultural resources contribute to its status as a key food producer in North Africa. Despite utilizing 30 million hectares of land for agriculture, Morocco still imports 25% of its cereals due to traditional farming methods and land ownership complexities. The sector faces water scarcity, climate change impacts, and smallholder farmer issues. Initiatives like the Green Morocco Plan and the Green Generation Strategy aim to enhance agricultural practices, water management, and rural development. Legislative measures supporting women's land rights promote gender equality in agriculture. International institutions like the AfDB and the World Bank support Morocco's agricultural development efforts. Various ministries are also actively working on strategies to manage drought-related losses and damages, including renewable energy programs, agricultural and water management plans, and infrastructure initiatives. Government initiatives like the Integrated Program for Drought Impact Mitigation (PRIDP) aim to address agricultural losses through financial assistance, water management, and asset protection. Morocco's response to drought impacts underscores the importance of international climate finance and national strategies to mitigate agricultural vulnerabilities and promote sustainability. The government has implemented strategic initiatives like the Green Morocco Plan and the Green

Generation Strategy to boost the agricultural sector's profitability and economic impact. These plans focus on improving regulatory frameworks, market access, and environmental sustainability. Programs such as PADIDZAR and the Competitive and Resilient Cereal Crop Development Support Program aim to enhance agricultural practices, water management, and rural development, aligning with the SDGs. Legislative measures have also been enacted to protect women's land rights, promoting gender equality in agriculture. International institutions like the AfDB and the World Bank have supported Morocco's agricultural development, adapting to changing sector needs. However, challenges for smallholder productivity remain due to climate change.

Finally, Chap. 12 traces the evolution and impact of environmental diplomacy over four decades through a systematic review of scholarly articles. Environmental diplomacy involves negotiating international agreements to address global environmental challenges. The literature landscape reveals a shift from bilateral to multilateral engagements, emphasizing the importance of cooperation and diplomacy in addressing issues like climate change and biodiversity loss. The chapter highlights the role of states, international organizations, and non-state actors in crafting and implementing environmental treaties. It discusses the challenges of power imbalances, implementation issues, and the positive impact of environmental diplomacy on raising awareness and fostering cooperation. Environmental diplomacy has evolved from a narrow conceptualization in the 1980s to a focus on internationalization in the 1990s, contextualizing shifts towards multilateral diplomacy in the 2000s, and advancing peace and equality considerations in the 2010s. The chapter also touches on economic impacts, the relationship between peace and the environment, and the importance of women's representation in global environmental governance.

## 3   The Way Forward in This Increasingly Complex World

This volume builds on two earlier volumes featuring research presented during NERPS 2022 (Sharifi et al. 2023a) and 2023 (Sharifi et al. 2023b) conferences. The first volume of these series was mainly focused on issues related to politics and governance, refugees, media, women, culture, and religion (Fig. 1.1b). In the second volume, the peace-sustainability nexus was explored with more attention to issues such as conflict resolution, peacebuilding, justice, youth and female empowerment, social dignity, and resource management, among other topics (Fig. 1.1a). This volume expands the coverage of the various factors enabling the pathways between peace and sustainability, with a more inclusive approach toward various dimensions of peace (both positive and negative peace) and sustainability compared to the previous volumes (Fig. 1.2). In particular, this volume puts more emphasis on environmental issues, institutional dimensions, and community-based efforts for navigating the challenges to peace and sustainability. Overall, it can be seen that, over time, the conference has been successful in expanding the boundaries of the nexus and exploring its multiple dimensions. This way, it has contributed to addressing multiple priorities outlined in Sharifi et al. (2021a, b).

Drawing from the insights presented in this volume, we outline some key recommendations for research and practice. First, global action must consider community-based efforts given the differentiated impacts of climate change and conflicts. These impacts are magnified at the local level and are especially detrimental to humans and nature in contexts where institutional capacities are weak to adapt to environmental issues and conflict risks. However, as several chapters in this volume have shown, there are existing local practices and structures that affected communities leverage in order to navigate these challenges. While a breadth of studies already contextualizes peace and sustainability at the local level, it remains important to consider how macro-level policymaking influences local dynamics and vice versa. Doing so provides a more balanced perspective and more relevant policy implementation.

**Fig. 1.1** The overall thematic focus of the second (**a**) and first (**b**) volumes, representing research presented during the first two NERPS conferences (reproduced from Sharifi et al. 2023b)

**Fig. 1.2** The overall thematic focus of this volume

Second and related to the first is the recognition of power asymmetries in both research and practice. For instance, some chapters in this volume demonstrate the vulnerability and resilience of Indigenous communities. While support for sustainable and peace-enhancing practices among Indigenous communities is crucial for developing policies that contribute to environmental justice and equitable climate action, researchers and practitioners alike must be careful not to instrumentalize, romanticize, and exhaust Indigenous agency. Research and policies despite good intentions may run the risk of exacerbating colonial legacies when they fail to account for historical injustice and structural forms of violence. It is, therefore, imperative to balance external support and local ownership to ensure just, inclusive, and equitable pathways for achieving peace and sustainability.

Third and to support the two recommendations is the promotion of transdisciplinary research and inter-stakeholder policy engagement. The concepts and issues covered in this volume draw on several disciplines and are relevant to multiple sectors. For example, sustainable agriculture and energy transition are not just a matter of economic development and environmental conservation, they also rest on social empowerment and distributional justice. Similarly, environmental peacebuilding and environmental diplomacy, informed by studies on resource management as well as politics and policy, need the contribution and cooperation of various stakeholders across levels for these efforts to be mutually beneficial to all stakeholders. As such, research on peace and sustainability must draw on inter- and transdisciplinary collaborations especially in this increasingly complex world.

# References

Arora P (2024) COP28: Ambitions, realities, and future. Environ Sustain 7(1):107–113. https://doi.org/10.1007/s42398-024-00304-0

Bailliet CM (2024) Peace and solidarity: dilemmas in an age of polarization and decoupling. In Bailliet CM (ed) Research handbook on international solidarity and the law. Edward Elgar Publishing, pp 80–100. https://doi.org/10.4337/9781803923758.00007

Berkes F (2018) Sacred ecology, 4th edn. Routledge

Berman N (2023, September 18) 2023: a summer of extreme weather. Council of Foreign Relations. https://www.cfr.org/article/weather-summer-2023-was-most-extreme-yet

Bin-Nashwan SA, Hassan MK, Muneeza A (2024) Russia-Ukraine conflict: 2030 Agenda for SDGs hangs in the balance. Int J Ethic Syst 40(1):3–16. https://doi.org/10.1108/IJOES-06-2022-0136

Candelaria JL, Sharifi A, Simangan D, Tabosa RMR (2023) A critical analysis of selected global sustainability assessment frameworks: Toward integrated approaches to peace and sustainability. World Dev Perspect 32:100539. https://doi.org/10.1016/j.wdp.2023.100539

Galtung J (1969) Violence, peace, and peace research. J Peace Res 6(3):167–191. https://doi.org/10.1177/002234336900600301

Galtung J (1990) Cultural violence. J Peace Res 27(3):291–305. https://doi.org/10.1177/0022343390027003005

Gledhill J, Bright J (2019) Studying peace and studying conflict: complementary or competing projects? J Glob Secur Stud 4(2):259–266. https://doi.org/10.1093/jogss/ogz001

Koga K (2023) Institutional dilemma: quad and ASEAN in the Indo-Pacific. Asian Perspect 47(1):27–48. https://doi.org/10.1353/apr.2023.0001

Poynting M, Rivault E (2024, January 9) 2023 confirmed as world's hottest year on record. https://www.bbc.com/news/science-environment-67861954

PRIO (2024, June 10) New data shows record number of armed conflicts. Peace Research Institute Oslo. https://www.prio.org/news/3532

Purvis B, Mao Y, Robinson D (2019) Three pillars of sustainability: in search of conceptual origins. Sustain Sci 14(3):681–695. https://doi.org/10.1007/s11625-018-0627-5

Redclift M (2005) Sustainable development (1987–2005): an oxymoron comes of age. Sustain Dev 13(4):212–227. https://doi.org/10.1002/sd.281

Richmond OP (2008) Peace in international relations. Routledge

Rockström J, Steffen W, Noone K, Persson Å, Chapin FS, Lambin EF, Lenton TM, Scheffer M, Folke C, Schellnhuber HJ, Nykvist B, de Wit CA, Hughes T, van der Leeuw S, Rodhe H, Sörlin S, Snyder PK, Costanza R, Svedin U et al (2009) A safe operating space for humanity. Nature 461(7263):472–475. https://doi.org/10.1038/461472a

Sala S, Ciuffo B, Nijkamp P (2015) A systemic framework for sustainability assessment. Ecol Econ 119:314–325. https://doi.org/10.1016/j.ecolecon.2015.09.015

Sharifi A, Simangan D, Kaneko S, Virji H (2021a) The sustainability–peace nexus: why is it important? Sustain Sci 16:1073–1077. https://doi.org/10.1007/s11625-021-00986-z

Sharifi A, Simangan D, Kaneko S (2021b) The literature landscape on peace–sustainability nexus: a scientometric analysis. Ambio 50(3):661–678. https://doi.org/10.1007/s13280-020-01388-8

Sharifi A, Simangan D, Kaneko S (2023a) Integrated approaches to peace and sustainability. Springer Nature Singapore. https://doi.org/10.1007/978-981-19-7295-9_1

Sharifi A, Simangan D, Kaneko S (2023b) Bridging peace and sustainability amidst global transformations (1st 2023. ed.). Springer Nature Singapore. Imprint: Springer

SIPRI (2023) SIPRI yearbook 2023: armaments, disarmament and international security. Oxford University Press

Stewart F (2009) Policies towards horizontal inequalities in post-conflict reconstruction. In Addison T, Brück T (eds) Making peace work: the challenges of social and economic reconstruction. Palgrave Macmillan UK, pp 136–174. https://doi.org/10.1057/9780230595194_7

Tidwell A (2001) Conflict resolved?: A critical assessment of conflict resolution. A&C Black

UN (2015, October 21) Transforming our world: the 2030 agenda for sustainable development. A/RES/70/1. https://sdgs.un.org/2030agenda

UN (2024a, June 28) World getting a 'failing grade' on Global Goals report card I UN News. UN News. https://news.un.org/en/story/2024/06/1151606

UN (2024b) The sustainable development goals report 2024. https://unstats.un.org/sdgs/report/2024/

WCED (1987) Our common future. World commission on environment and development. https://sustainabledevelopment.un.org/content/documents/5987our-common-future.pdf

# Environmental Ethics in Islamic Teachings: Discussing Ethical Principles in Islamic Teachings that Emphasize Environmental Protection and Preservation

Aji Kurbiyanto, Desfa Yusmaliana, Fifin Fitriana, Eka Altiarika, and Fadillah Sabri

**Abstract** Islamic ethics offers valuable guidance for addressing pressing environmental issues. Ethical responsibility in Islamic teachings, especially for protecting and preserving the environment, is one of the core principles of Islam. The Qur'an emphasizes that humans are servants of the earth with a duty to safeguard the environment. Hadith and Sunnah provide ethical insights that encourage conservation and compassion. This article employs a systematic literature review methodology, examining Islamic ethical principles related to environmental care and synthesizing the findings from various sources. It reveals a deep interconnectedness among all living things. Exploring concepts such as Tawhid (oneness of God) and Ihsan (excellence in conduct) in environmental management provides a broad perspective on the ethical obligations of Muslims towards the environment. These principles remain important amid contemporary environmental challenges, offering valuable guidance for addressing pressing environmental issues.

**Keywords** Environment · Islamic environmental ethics · Islamic teachings · Tawhid and Ihsan

## 1 Introduction

The environmental crisis has become the most significant problem in human life. Various forms of pollution have had and will continue to have, a profound impact on future generations. This impact has naturally attracted the attention of many parties. For example, efforts through education (Fien 1995; Michelsen and Fischer

A. Kurbiyanto · D. Yusmaliana (✉) · F. Fitriana · E. Altiarika · F. Sabri
KSDA, Universitas Muhammadiyah Bangka Belitung, Kota Pangkal Pinang, Indonesia
e-mail: desfa.yusmaliana@unmuhbabel.ac.id

D. Yusmaliana
SHBIE, Universiti Brunei Darussalam, Bandar Seri Begawan, Brunei Darussalam

A. Sharifi et al. (eds.), *Navigating Peace and Sustainability in an Increasingly Complex World*, World Sustainability Series, https://doi.org/10.1007/978-981-97-8772-2_2

15

2017) began in the late 1960s with environmental education (EE) initiatives aimed at raising awareness about pollution affecting water, soil, and air (Wals and Benavot 2017). Traditionally, EE has focused on providing knowledge about how and why we should behave in environmentally friendly ways (Krettenauer 2017). Furthermore, synergy with environmental considerations has become essential in the social sector, particularly village development (Siombo and Adi 2021).

However, these efforts must be complemented by other factors, such as an individual's emotional connection with nature (Braun and Dierkes 2017; Collado et al. 2020; Otto et al. 2019), which can arise from human connections with the natural world (Makoul 2022) and practices derived from religious teachings. Tanzeh and Junaris (2019) revealed that religious teachings act as managers or controllers of behaviors and actions born from desires based on emotions. If religious education is integrated into daily life early, behaviors will be more controlled (Hadayani et al. 2021; Marcus and McCullough 2021; Siddique 2024; Sukino and Utami 2020; Syafi'i and Mardiyah 2023; Yulianingsih 2023). Therefore, in the face of the environmental crisis, it is not enough to merely provide or transfer knowledge; understanding and appreciating the importance of environmental ethics in Islamic teachings is crucial to overcoming contemporary challenges that threaten the balance of our planet. For environmental conservation, Islam emphasizes the need for specific guidelines such as ecological philosophy, Sufism ecology, and eco-Usul al-fiqh, which are interpretations of the general messages of the primary sources of Islamic teachings, the Qur'an and Hadith (Najib et al. 2020).

The issue statement underscores the urgency of integrating ethical considerations from Islamic teachings into broader discussions about environmental sustainability. It encourages critical reflection on how ethical principles rooted in faith can contribute meaningfully to the discourse around environmental conservation. As we navigate the complexities of the modern world, environmental degradation has become an urgent global concern. In this context, articulating problems within the framework of Islamic environmental ethics becomes imperative. By highlighting the importance of addressing environmental issues from the Islamic ethics perspective, Islamic teachings offer unique insights into encouraging sustainable practices and responsible management, such as emphasizing hygiene, a clean lifestyle, and avoiding redundancy (Omar et al. 2018).

Therefore, this study aims to investigate the rich ethical principles in Islamic teachings that specifically emphasize environmental protection and preservation. Using a comprehensive literature review methodology, drawing insights from the Qur'an, Hadith, Sunnah, and broader Islamic literature, the study sheds light on the profound teachings guiding Muslims toward ecological stewardship.

## 2 Literature Review

The role of religious teachings in enhancing environmental management has been the subject of several systematic literature reviews (SLR) in recent years. These studies highlight the significant influence of religious doctrines on environmental ethics and practices. For instance, in Sayem (2019), Cobb attempts to address the ecological crisis from the perspective of beliefs found in Christianity, and Nasr addresses ecological issues from the Islamic understanding of nature. Similarly, Sadowski and Ayvaz (2023) found in their surprising analysis that both the Bible and the Qur'an encourage their followers to build harmonious relationships with nature. Therefore, efforts to improve environmental sustainability can be mobilized through communities, particularly among the younger generation who believe in God (Ives et al. 2023; Körtner 2022; Lowe et al. 2023).

Another example of comprehensive SLR studies is linked to various scopes and objectives. Al-Jayyousi et al. (2022) examined multiple models, practices, and drivers of sustainable Islamic development to inform a new discourse on the potential implementation of Islamic development models that can promote inclusive, pro-poor, and human-centered development. Similarly, a review conducted by Aziz et al. (2018) focused on the importance of considering cultural and religious factors, such as Islamic Work Ethics (IWE), in understanding and encouraging pro-environmental behavior among employees. Furthermore, Ab Manaf and Ahmad Rashid (2022), using two leading databases, Scopus and Web of Science, and one supporting database, Google Scholar, from 2015 to 2022, still found that the level of sustainable thinking among students was still low, and the implementation of environmental education was unsatisfactory.

Although individual studies have addressed specific aspects of Islamic environmental understanding, such as principles derived from the Qur'an and Hadith that encourage conservation and sustainability and emphasize human responsibility to act as guardians of the Earth (Gueye and Mohamed 2022; Hayat et al. 2023; Mangka et al. 2022; Mustolikh et al. 2022; Roy Purwanto et al. 2022; Yusmaliana et al. 2023), there remains a significant gap in the systematic exploration of Islamic teachings related to environmental ethics itself, which is fundamental to its application. This gap highlights the need for an SLR focused on the intent of environmental ethics in various studies, including from an Islamic perspective, and the correlation of Islamic environmental ethics in contemporary contexts.

## 3 Methodology

The systematic literature review (SLR) method is used in this article. The choice to conduct a systematic literature review was made because the systematic aspects of the review can act as a foundation for future empirical research. The results of this systematic review may form the basis for empirical studies that will examine the proposed relationships. This literature review follows the guidelines described in PRISMA (Moher et al. 2009; Perry and Hammond 2002; Petticrew and Roberts 2008). Systematic reviews and meta-analyses are becoming increasingly important in developing various studies and ensuring the justification of subsequent research. The following is the research flow of the Systematic Literature Review (SLR) method as described by Perry and Hammond (2002) which can be seen in the table below:

In the search process, researchers entered the phrase "Islamic environmental ethics" using the Publish or Perish (PoP) application with a publication period of the past five years. The search results were sorted by examining the papers' titles, abstracts, and content. After this process, the primary study results obtained were used as source material for analysis related to the research theme. Data analysis was conducted using content analysis (Moleong 2006), emphasizing intertextuality. The literature in the field of environmental ethics was corroborated by literature in the field of Islamic studies.

Using the Publish or Perish (PoP) application, data was collected from the Crossref website to facilitate finding the necessary articles. The keyword used in the search was "Islamic environmental ethics," which resulted in 1000 papers published between 2018 and 2023.

The researchers then carried out the review process by capturing only publications in the form of full-paper articles taken for review. Of the 1000 papers initially grouped, 525 articles passed the initial screening, excluding book chapters, monographs, proceedings, reference entries, and other forms. The researchers then filtered these 525 articles for compatibility with the theme, including only those with full access for further study. The exclusion criteria were: (1) articles not in English or Indonesian and (2) articles not related to environmental ethics or environmental ethics from an Islamic perspective.

This process resulted in 43 articles that met the appropriate titles and exclusion criteria. Upon further review, the researchers downloaded the full articles and found only 23 full papers. After a comprehensive reading, only 19 articles were selected for detailed study based on their content. The selection process was conducted to ensure that the final collection of articles for in-depth analysis were published in either Indonesian or English, were full-paper articles that fit the research issue, and were available in full text for comprehensive review. Furthermore, the articles had to be of sufficient quality and significance as determined by a thorough reading, and they had to be pertinent to environmental ethics or environmental ethics from an Islamic perspective.

The following outlines the literature search process related to the chosen theme (Fig. 1).

After completing the article selection procedure, the results of the main study are utilized for additional analysis, as shown in Table 1.

As shown in Table 2, based on the article screening data, the selected articles were published as follows: 3 articles in 2018, 5 articles in 2019, 2 articles in 2020, 3 articles in 2021, 3 articles in 2022, and 3 articles in 2023. They were published by domestic and international research journals and authored by researchers from various countries—including Indonesia, Malaysia, the Netherlands, Pakistan, Colombia, the

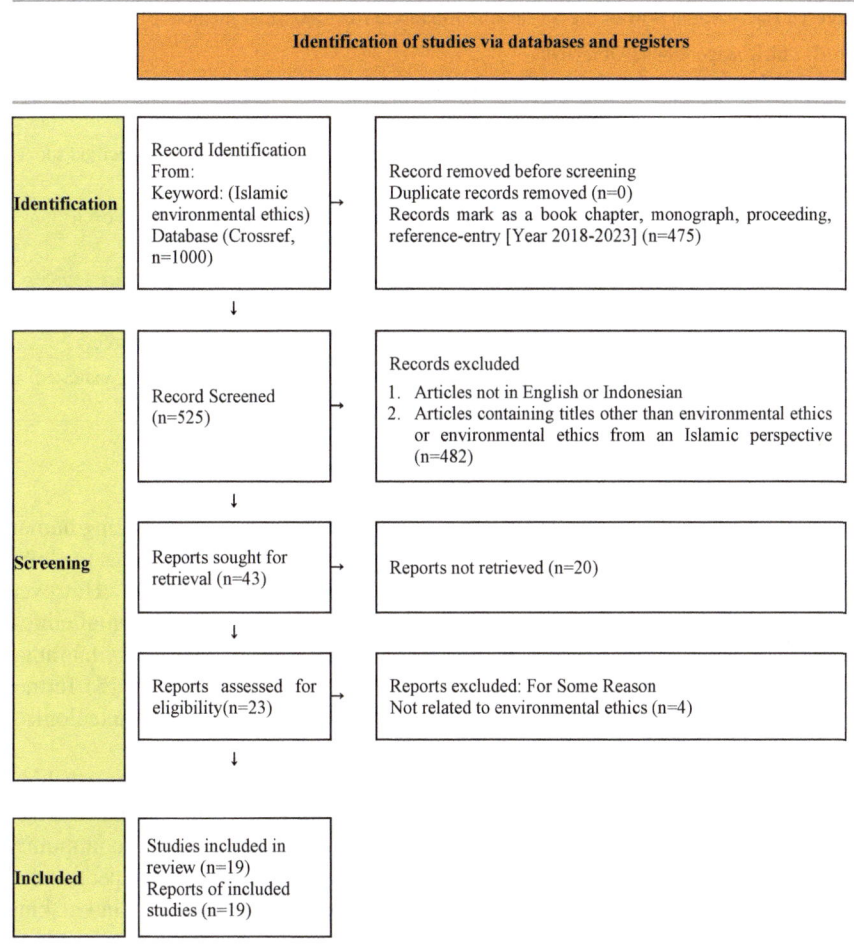

**Fig. 1** Prisma: identification of studies via databases and registers. *Source* Page et al. (2021), visit: http://www.prisma-statement.org/

United Kingdom, the United States, Germany, and Australia. Ultimately, they discuss and examine environmental ethics relevant to the research topic (Fig. 2).

In the 19 search result articles, researchers grouped the results into three, namely (1) related to environmental ethics, (2) key sources such as the Qur'an, Hadith, Sunnah, and other Islamic literature related to environmental ethics, and (3) the correlation of Islamic environmental ethics in the contemporary context.

1. Environmental ethics

This article group discusses the concept of environmental ethics, where the scope involves various related ethical frameworks. Dzwonkowska (2018) combines virtue ethics with environmental ethics to form Environmental Virtue Ethics (EVE).

**Table 1** The research flow of the Systematic Literature Review (SLR) method

| No. | SLR stage flow | Objective |
| --- | --- | --- |
| 1 | Scope determination | Identifying and determining the scope of research topics to be reviewed |
| 2 | Literature search | Carry out relevant literature searches according to the specified scope |
| 3 | Literature selection | Selecting literature based on predetermined inclusion and exclusion criteria |
| 4 | Literature evaluation | Evaluate the reliability and validity of the selected literature sources |
| 5 | Data extraction | Extract data from each article or relevant information source |
| 6 | Analysis and interpretation | Synthesize data from various literature sources to identify patterns, trends, or main findings relevant to the research topic |

*Source* Perry and Hammond (2002)

Through virtue ethics, which contains an anthropocentric view emphasizing human interests and welfare above those of other creatures or the environment as a whole, various controversies arise due to its perceived proximity to egoism. However, Deplazes-Zemp (2023) revealed that in non-anthropocentric environmental ethics, the term "value" usually addresses why natural entities are relevant and vital rather than why they are desirable or worth fighting for. Dzwonkowska (2018) further explains EVE from the perspectives of naturalistic, pluralistic, and teleological concepts by Ronald Sandler.

Contrary to Dzwonkowska's view that virtue ethics leads to environmental ethics, Makoul (2022) argues that the platform for promoting environmental ethics arises from the reciprocal relationship between lawmaking and the formation of community virtue values or ethics. Although this view contrasts with the common perspective that virtue ethics influence the creation of law ethics, Aristotle contended that lawmaking can create a holistic platform for people to learn how to practice environmental ethics, which leads to the passage of new environmental regulations and protection laws.

2. Key sources such as the Qur'an, Hadith, Sunnah, and other Islamic literature related to environmental ethics
3. Several previous studies have examined this aspect of ethics in the context of the environment, but the majority are limited to philosophical perspectives and personal opinions. The clarification of environmental ethics will be explored in the context of Islamic teachings, which have a recognized foundation and are considered a mercy for all nature, using the Qur'an as its main footing. Previously, in sorting literature, ten articles discussing Islamic environmental ethics had been identified, including works such as Amri and Adawiyah Mohd (2022), Aziz et al. (2018), Gulzar et al. (2021), Hayat et al. (2023), Hummel and Daassa (2019), Irawan et al. (2021), Islam et al. (2021), Mauluah et al. (2021), Omar et al. (2018), Sayem (2019). The correlation of environmental ethics Islam in the contemporary context.

**Table 2** List of selected articles

| No. | Title | Author/s | Year | Journal |
|---|---|---|---|---|
| 1 | Islamic ethics of waste management towards sustainable environmental health | Omar, Suhaila Mohd | 2018 | International Medical Journal Malaysia |
| | | Chowdhury, Ahmed Jalal Khan | | |
| | | Hashi, Abdurezak Abdulahi | | |
| 2 | Linking Islamic work ethics and pro-environmental behaviour: a systematic review | Aziz, Faiq | 2018 | The Journal of Social Sciences Research |
| | | Mahadi, Nomahaza | | |
| | | Mohammad, Jihad | | |
| 3 | Is environmental virtue ethics anthropocentric? | Dzwonkowska, Dominika | 2018 | Journal of Agricultural and Environmental Ethics |
| 4 | The goodness of means: instrumental and relational values, causation, and environmental policies | Baard, Patrick | 2019 | Journal of Agricultural and Environmental Ethics |
| 5 | From environmental stewardship to environmental holiness: the evolution of methodist environmental witness, with a focus on climate change | Stephens, Darryl W | 2019 | Journal of Religious Ethics |
| 6 | "Walking gently on the earth": an interview with nana firman on islamic environmental ethics | Hummel, Daniel | 2019 | The Journal of Islamic Faith and Practice |
| | | Daassa, Mohamed | | |
| 7 | Exemplarist environmental ethics: Thoreau's political ascetism against solution thinking | Balthrop-Lewis, Alda | 2019 | Journal of Religious Ethics |
| 8 | Religions and environmental ethics | Sayem, Md Abu | 2019 | Australian Journal of Islamic Studies |
| 9 | Environmental conservation paradigm in Islamic ecological perspective | Najib, Moh | 2020 | International Journal of Psychosocial Rehabilitation |
| | | Saefullah, Ujang | | |
| | | Haryanti, Erni | | |
| | | Haetami, Indian | | |
| | | Maryati, Delis Sri | | |
| 10 | Ethico-Religious Green Supply Chain Management (GSCM): embedding Islamic ethics' codes for improving environmental concerns | Islam, Md Shamimul | 2021 | Journal of Islamic Accounting and Business Research |
| | | Karia, Noorliza | | |
| | | Taib, Fauziah Md | | |
| | | Ara, Husna | | |
| | | Moeinzadeh, Soroush | | |

(continued)

**Table 2** (continued)

| No. | Title | Author/s | Year | Journal |
|-----|-------|----------|------|---------|
| 11 | Environmental ethics towards sustainable development in Islamic perspective | Gulzar, Adil | 2021 | Ethnobotany Research and Applications |
| | | Islam, Tajamul | | |
| | | Khan, Muhammad Anees | | |
| | | Haq, Shiekh Marifatul | | |
| 12 | Design of elementary/MI mathematics worksheets based on islamic environmental ethics | Muluah, Luluk Marsigit Wangid, Muhammad Nur | 2021 | Muallimuna: Jurnal Madrasah Ibtiadiyah |
| 13 | Applying Ibn ʿArabī's Concept of Tajallī: a Sufi approach to environmental ethics | Irawan, Bambang | 2021 | Teosofia: Indonesian Journal of Islamic Mysticism |
| | | Nasution, Ismail Fahmi Arrauf | | |
| | | Coleman, Hywel | | |
| 14 | A review on Islamic environmental ethics: a solution microplastics-based water | Aini Ameera Muhammad Amri, Siti | 2022 | Journal of Islam and Contemporary Society |
| | | Adawiyah Mohd, Robiatul | | |
| 15 | Environmental ethics and environmental law: a virtuous circle | Makoul, Zoe | 2022 | Columbia Journal of Environmental Law |
| 16 | Bioethical analysis of sanitary engineering: a critical assessment of the profession at the crossroads of environmental and public health ethics | Eterović, Igor | 2022 | Ethics in Science and Environmental Politics |
| | | Buterin, Toni | | |
| | | Cinintya Pratama, Bima | | |
| | | Fakhruddin, Iwan | | |
| | | Wibowo, Hardiyanto | | |
| 17 | Artificial intelligence needs environmental ethics | Baum, Seth D | 2023 | Ethics, Policy and Environment |
| | | Owe, Andrea | | |
| 18 | The role of Islamic environmental ethics in the alleviation of climate challenges and the preservation of ecosystem | Hayat, Imran | 2023 | Russian Law Journal |
| | | Malik, Muhammad Sajad | | |
| | | Ali, Muhammad Waris | | |
| | | Husnain, Muhammad | | |
| | | Sharif, Muhammad | | |

(continued)

**Table 2** (continued)

| No. | Title | Author/s | Year | Journal |
|---|---|---|---|---|
| | | Haleem, Abdul | | |
| 19 | Beyond intrinsic and instrumental: third-category value in environmental ethics and environmental policy | Deplazes-Zemp, Anna | 2023 | Ethics, Policy and Environment |

**Distribution of Articles Based on Journal Publishing Country**

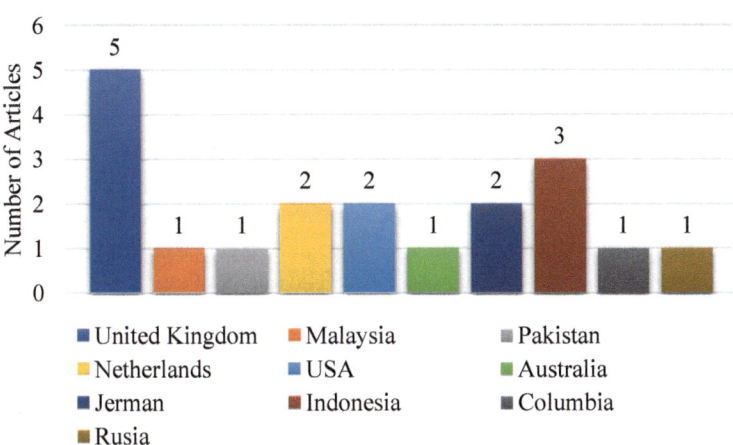

**Fig. 2** Articles distribution

In the context of government, lawmaking can shape environmental awareness through education, especially by fostering community connections with nature through projects such as creating national parks and allocating public land (Makoul 2022). Professions such as sanitary engineering also contribute to this effort by improving individual and community health, preventing disease, controlling hazardous actions by following laws and regulations, and promoting public education (Eterović and Buterin 2022) based on the premise that environmental ethics are an integral part of professional ethics (Des Jardins 2013).

Even in the context of rapid technological advancement, the emergence of artificial intelligence (AI) has introduced new dimensions to environmental ethical issues. Baum and Owe (2023) highlight that environmental ethicists can examine the environmental impact of AI, such as its energy footprint and potential environmental protection applications. Additionally, environmental ethicists can evaluate the ethics of new situations enabled by AI, such as computer-based artificial life and artificial ecosystems. While AI can assist in solving environmental problems, it also poses significant environmental challenges. For example, a study found that training one AI model can produce greenhouse gas emissions comparable to the total lifetime emissions of one or several cars (Strubell et al. 2020).

Much can be seen in the correlation between environmental ethics and the current context. In Islam, the discussion of water and corruption is particularly significant. Corruption in Islam is not limited to human actions that violate, discriminate, or manipulate others; it also encompasses activities that negatively impact other organisms, including animals and plants. Corruption thus includes, among others, deforestation, the disposal of toxic waste, and the use of synthetic chemicals (Ashtankar 2016).

## 4 Findings and Discussion

### 4.1 Environmental Ethics

Ethics is a systematic framework for evaluating behavior from the standpoint of good or bad, right or wrong. It focuses on a connected character, a high-quality approach, acting satisfactorily, following rules and laws, responsibility and ability, fair and reasonable accountability, and more. The ancient Greeks argued that moral character could be developed through education to achieve fulfillment through proper behavior and pure thinking. This concept, known as ethics, originates from the Greek word "ethos." Jackson (2019) revealed that several models of virtue and ethics were developed by philosophers in the fifth century BC to guide character formation. Today's ethical standards are stringent and prescriptive in many fields.

The study of ethical issues raised by human interaction with the environment is known as environmental ethics. It is essential to understand that ethical statements are prescriptive, not descriptive or predictive; they are aspirational and normative (Palmer et al. 2014). For example, the statement that society should reduce ecological impacts is prescriptive. While it may seem that today's lifestyles are unlikely to change soon, these claims remain true as aspirational goals. Therefore, prescriptive statements cannot be reduced to descriptive claims about individual behavior and beliefs or predictions about potential future events. Instead, they outline the traits, customs, and behaviors that must be achieved, even if they are challenging to fulfill.

The development of environmental ethics has become a concern for various religions. In the late 1990s, Christianity identified three fundamental environmental ethics: stewardship, environmental justice (specifically focusing on all forms of disparities), and environmental spirituality (emphasizing a return to nature). The cultural aspects of religion are intertwined with the structure of religious environmental movements (Ellingson et al. 2012). Similarly, the Islamic community experienced a resurgence in environmental ethics in the late 1990s. Muslims in the West were the first group to articulate modern environmental ethics in Islam based solely on the Qur'an and Sunnah. European protest movements in the 1960s and 70s, which opposed mass consumption, impersonal materialism, and overproduction, influenced this resurgence (Ragozina 2023). Verses forbidding the plunder of the earth (Qur'an 7:85) and emphasizing the prohibition of spreading corruption on the earth (Qur'an

28:77) are among the key references. The Qur'anic scholar Ibn Kathir (d. 1373) interprets verse 28:77 to mean that believers should "not allow their purpose of spreading corruption on the face of the earth and corrupting God's creation" (Hummel and Daassa 2019).

In principle, it is recognized that every religion has regulated all aspects of human life, including interaction with nature. Human relations with nature must be built comprehensively, starting from human behavior towards the environment. In reality, the relationship between humans and nature must be established harmoniously to achieve sustainability of life (Kar and Tripathy 2022). Christianity advocates responsible management of the Earth (Pasaribu et al. 2022; Twardziłowski 2020). Hinduism promotes a deep sense of connectedness with nature (Ahrens 2022; Sen 2021). Buddhism emphasizes the principles of interdependence and impermanence, encouraging ethical behavior toward the environment (Kontio 2020; Lim 2019). All these religions recognize the importance of fostering ethical and harmonious relationships between humans and nature.

## 4.2 Islamic Ethical Principles for Environmental Protection and Preservation in the Perspective of the Qur'an and Hadith

Islamic ethical principles for environmental protection and preservation, as depicted in the Quran and Hadith, emphasize the harmonious relationship between humans and nature. The Quran describes nature as a sign of Allah's power, inviting humans to reflect on and take responsibility for the environment. For example, QS. An-Nahl: 10 highlights the importance of rainwater as a source of life for humans and plants and reminds us of our responsibility to care for and protect the environment. Similarly, in the hadiths, the Prophet (صلى الله عليه وسلم) underlined the material and spiritual benefits of treating the environment well. As conveyed in the following hadith:

> Narrated Anas bin Malik: The Prophet (صلى الله عليه وسلم) said, "No one among Muslims plants a tree or sows a seed, and then a bird, or a person or animal eats it, but it is considered a charitable gift to him." (*Sahih Al-Bukhari 2320, Book 41, Hadith 1*, n.d.)

For this reason, the Islamic perspective emphasizes that everything in nature, other than Allah, needs to be valued and kept in balance (Muhaimin 2015). The Qur'an, as the central religious text in Islam, provides fundamental guidance on ethical principles for environmental protection and preservation. A key aspect highlighted in the Qur'anic verses is the great responsibility given to humans as stewards of the earth. The Qur'an emphasizes the concept of guardianship, describing human beings as custodians entrusted with the care of the environment (Hayat et al. 2023). Verses like "It is He who has made you successors (Caliphs) on Earth…" (Qur'an 35:39) underline the Qur'anic perspective that positions humans not just as inhabitants but as guardians with a moral obligation to preserve nature. Islamic environmental ethics finds its epistemological foundation in these three precepts of fiqh-based ethics—*maslahah mursalah* (public interest), *mizan* (balance), and *amanah* (trustworthy).

Furthermore, the teachings of the Qur'an investigate the interconnectedness of all living things, emphasizing the intricate web of relationships in ecosystems. The verses highlighting the interdependence of creatures and the balance of nature reinforce the idea that environmental care is an important aspect of ethical behavior in Islam. For example, the Qur'an states, "And there is no creature on earth or a bird that flies with its wings except that they are communities like you..." (Qur'an 6:38). This interconnected worldview serves as the basis for environmental ethics in Islam, urging believers to recognize the value of all living entities and to promote responsible stewardship.

Cultivating an understanding of the environment presents more significant challenges than simply providing knowledge about environmental issues, which tend to be the focus of many environmental education programs. Teaching environmental understanding in Islam requires an emphasis on active involvement during the process. This approach must integrate Islamic principles and teachings, ensuring that environment-related decision-making and actions align with genuine environmental ethics.

One example that highlights the importance of environmental awareness in Islamic teachings can be found in Abdullah ibn Amr's statement:

حَدَّثَنَا مُحَمَّدُ بْنُ يَحْيَى، حَدَّثَنَا قُتَيْبَةُ، حَدَّثَنَا ابْنُ لَهِيعَةَ، عَنْ حُيَيِّ بْنِ عَبْدِ اللَّهِ الْمَعَافِرِيِّ، عَنْ أَبِي عَبْدِ الرَّحْمَنِ الْحُبُلِيِّ، عَنْ عَبْدِ اللَّهِ بْنِ عَمْرٍو، أَنَّ رَسُولَ اللَّهِ ـ صلى الله عليه وسلم ـ مَرَّ بِسَعْدٍ وَهُوَ يَتَوَضَّأُ فَقَالَ " مَا هَذَا السَّرَفُ " . فَقَالَ أَفِي الْوُضُوءِ إِسْرَافٌ قَالَ " نَعَمْ وَإِنْ كُنْتَ عَلَى نَهَرٍ جَارٍ " .

It was narrated from 'Abdullah bin 'Amr that: The Messenger of Allah passed by Sa'd when he was performing ablution, and he said: "What is this extravagance?" He said: "Can there be any extravagance in ablution?" He said: "Yes, even if you are on the bank of a flowing river." (*Sunan Ibn Majah 425, Book 1, Hadith 159*, n.d.)

This hadith underlines the importance of practicing moderation and avoiding waste, even in seemingly small acts such as ablution. It reminds individuals of their responsibility to conserve and value natural resources, including water, as a reflection of their faith and Islamic teachings (Bsoul et al. 2022).

The ethical principles that guide environmental conservation in Islam extend beyond the Qur'an to include the Hadith (sayings and actions of the Prophet Muhammad) and the Sunnah (the traditions of the Prophet). These sources offer numerous ethical insights that further enrich the understanding of Muslims' responsibility towards the environment. The hadith serves as a practical guide, providing specific examples and stories that illustrate the importance of environmental protection in the Islamic tradition (Ali and Gul 2018).

Manoiu et al. (2016) and Mauluah et al. (2021) have highlighted several principles of Islamic environmental ethics, particularly regarding the treatment of water, animals, plants, earth, and air. The Qur'an affirms the importance of water as a source of life (21:30, 25:54, 24:45), repeatedly mentioning the significance of water resources and river flows and teaching that water is a gift that must be appropriately maintained. Regarding animals, the Qur'an emphasizes their meaningful creation for human needs (16:5, 16:7, 16:8, 22), provided as food sources and for other purposes. The relationship between humans and plants is also highlighted, with the Qur'an emphasizing sustainable use (QS An-Nahl:114) and the Prophet's hadiths advocating for planting trees and forbidding their unnecessary cutting. Islam entrusts humans as caliphs, or stewards, of the earth, assigning them the responsibility of safeguarding and prospering the planet (QS Al-Baqarah [2]:30, QS An-Nur [24]:55). Additionally, Islamic environmental ethics emphasize the preservation of air, focusing on maintaining water availability due to its connection with the water cycle, as noted in the Qur'an's mention of how God sends wind to bring rain (56:68–70, 7:57).Ultimately, Islamic environmental ethical principles guide environmental conservation, and sources such as the Qur'an and Hadith provide insights that further enrich Muslims' understanding of their responsibilities toward the environment. Through this understanding, Muslims are empowered to practice environmental ethics daily, creating harmony between humans and nature.

## 4.3 Interconnection in Islamic Environmental Ethics

### Exploration of the Concept of Tawhid (Oneness of God) in Relation to Environmental Management

The essence of Islamic environmental ethics lies in the profound concept of Tawhid, emphasizing the oneness of Allah. Tawhid serves as a fundamental principle shaping the Islamic worldview and, in the context of environmental management, underscores the interconnectedness of all existence (Irawan et al. 2021; Najib et al. 2020). The belief in the oneness of Allah goes beyond theological considerations to have practical implications on how Muslims approach environmental management.

Many researchers agree that environmental management is closely related to the concept of Tawhid. For instance, Gulzar et al. (2021) revealed that the Islamic perspective on the environment acknowledges that Allah is the creator and designer of everything in the universe, including the surroundings. This view is based on the Qur'an, which describes how Allah adorns the heavens and the earth with various elements of nature (Qur'an, 41:53).

The concept of Tawhid, which is part of the environmental ethics dimension, highlights the importance of humans recognizing the omnipotence of God and building strong faith without dichotomizing between religious and scientific views but complementing each other (Aini Ameera Muhammad Amri and Adawiyah Mohd 2022). Conversely, a lack of true faith in the One Almighty God who created humans and the environment leads to imbalance and a lack of harmony between humans and the environment.

### Discussion of Ihsan (Excellence or Benevolence) and Its Role in Shaping Muslim Ethical Obligations to the Environment

An integral component of Islamic environmental ethics is the concept of Ihsan, often translated as excellence or virtue. Ihsan is a multidimensional concept that permeates various aspects of a Muslim's life, including their ethical responsibility (Aini Ameera Muhammad Amri and Adawiyah Mohd 2022). Rooted in the teachings of the Prophet Muhammad, who defined Ihsan as "worshiping Allah as if you saw Him, and if you do not see Him, then [knowing] that He sees you," Ihsan encourages Muslims to act with the highest excellence and goodness in all aspects of their lives, including their relationship with the environment (Hadith, Sahih al-Bukhari).

Irawan et al. (2021) adopted Al Farabi's concept of tajalli and its implications for environmental ethics. Tajallī encourages man to see himself as a reflection of nature. One can enter into the inner meaning of nature by exploring the inner depths of oneself. On the contrary, another message from Tajalli is that the current ecological crisis is mainly caused by the neglect of the spiritual center, which is closely related to the ability to remember God and see Him anywhere.

The superiority of the Ihsan concept in forming a Muslim who cares about the environment is inseparable from the concept of Tawhid explained earlier. In Islam, Ihsan is known as high-performance work values (Wahab and Ismail 2019). In the context of environmental management, Ihsan manifests a commitment to caring for the environment beyond the minimum requirements. It calls on Muslims to approach environmental responsibility with a sense of superiority, going beyond what is only allowed to what is commendable (Watsiqotul et al. 2018). It could involve practices such as water conservation, waste reduction, and sustainable resource management. The Prophet Muhammad, through his actions and teachings, exemplified Ihsan in his treatment of nature, emphasizing the ethical imperative to act with benevolence towards all creation. In everyday life, the concept of Ihsan needs to be instilled from an early age, starting with habituation and direct experience. Mauluah et al. (2021) revealed that introducing the concept of Islamic environmental ethics simply began with the example of parents to children in the family environment, such as planting trees, caring for them, and using them sufficiently.

By discussing Ihsan's role in shaping Muslims' ethical obligations towards the environment, it should be underlined that the concept encourages a reflective and conscientious approach, encourages believers to pay attention to the impact of their actions on the environment, and inspires them to strive for the highest standards of ethical behavior.

## 4.4 Relevance of Islamic Environmental Ethics in Contemporary Context

In the contemporary context, the use of science and technology has grown more and more rapidly. However, greedy humans exploit the natural environment for economic prosperity. In other words, Sayem (2019) revealed that science and technology have given broad powers to demonstrate human mastery of nature but cannot provide strong ethical or moral awareness of how to utilize these powers to benefit all beings, including humans. It is undeniable that in recent decades, conventional ethical guidelines have indeed played an important role in developing new fields of science but do not control future technologies, so threats and pollution persist.

As we face the challenges of an increasingly global environment, Islamic ethical principles offer a valuable and enduring framework that remains highly relevant in addressing contemporary issues. Principles derived from Islamic teachings, including Tawhid and Ihsan, provide deep insight into ethical behavior that transcends temporal and geographical boundaries. In practical terms, cultivating an understanding of the environment within the framework of Islamic education involves adopting a prudent sequence of operations: first, reject; then, reduce, reuse; and finally, recycle. Through understanding the environment that shapes individuals' perceptions of the world, they will prioritize the use of reusable water bottles over single-use plastic bottles. Similarly, they will observe less paper waste in recycling bins, not because it has been dumped there, but because they consciously reduce paper consumption. In line with this thinking, Aziz et al. (2018) revealed that Islamic thinking about the role of such values will be able to help public and private sector organizations to more successfully communicate messages of environmental protection, encourage pro-environmental actions and contribute to improving the quality of life.

All actions related to caring for the environment are a shared responsibility. Various practices related to environmental protection are even collected in an association, both globally and locally. For example, the World Wide Fund for Nature (WWF), formerly called the World Wildlife Fund and still the official name in Canada and the United States, is an international non-governmental organization that has dealt with issues concerning conservation, research, and environmental restoration (Chairunnisa 2018). Similarly, the United Methodist Church's (UMC) involvement in various environmental initiatives, such as the "Climate Justice Now" movement in the Philippines, the establishment of the Climate Change Task Force in Liberia, and leadership in ecumenical efforts in Europe are interrelated in analyzing resolutions and topics related to environmental management, shared heritage, and the interlinked impacts of agricultural, technological, and military activities on the environment (Stephens 2019). In Cobb's thinking, as an environmental ethicist, it is stated that the Bible not only tells about human supremacy over non-human animals and plants but also makes them the custodians of God's creation (Sayem 2019).

Sutarto (2023) emphasizes that the internalization of Islamic values can be applied, for example, to clean living in the school environment through three patterns. First, recognition and understanding are done through the learning process, posters, and so on. Second, curriculum interventions are implemented by integrating them into subjects and assigning assignments. Third, habituation. Students are accustomed to behave and behave in caring for the environment. This pattern is carried out through several activities, such as habituating clean living and caring for the environment, planting, and giving rewards and punishments.

By conducting this analysis, this article seeks to highlight the enduring relevance of Islamic ethical principles in shaping responses to contemporary environmental challenges. This study underscores the potential of these principles to contribute meaningfully to global efforts aimed at environmental sustainability. As the world grapples with pressing environmental issues, ethical insights from Islamic teachings offer an enduring guide to ethical behavior, promoting harmonious coexistence between humans and the environment.

## 5  Conclusion

To put it briefly, several sources, including the Qur'an, Hadith, and broader Islamic literature, have provided significant insights into Islamic environmental ethics. Tawhid's principles emphasize the interconnectedness of all living things, and Ihsan, promoting excellence and virtue, provides a robust framework for environmental stewardship. These principles guide individual behavior and offer valuable perspectives for addressing contemporary environmental challenges on a global scale.

Islamic ethical principles are permanent and adaptable, making them crucial for environmental protection. Tawhid encourages a holistic approach to stewardship, while Ihsan calls for conscientious and proactive engagement with environmental issues. By integrating these principles into policies, practices, and lifestyles, individuals and communities can contribute to a more sustainable relationship with the environment.

Future research could explore how modern technology can support the application of Islamic environmental principles, including using digital platforms for education and awareness campaigns, as well as the implementation of environmentally friendly technologies. Comparative studies with other religious environmental ethics, practical integration into policies, empirical assessments, interdisciplinary approaches, and international collaborations can further expand the understanding and application of Islamic environmental ethics. These efforts will help foster a more rigorous and sustainable approach to environmental management globally.

# References

Ab Manaf NA, Ahmad Rashid R-A (2022) Systematic literature review: students' sustainable thinking toward environment. Malays J Soc Sci Human (MJSSH) 7(11):e001874. https://doi.org/10.47405/mjssh.v7i11.1874

Ahrens LS (2022) Know this to be the enemy: desire in the Bhagavad Gita and environmental destruction (Thesis) (Issue May). The State University of New Jersey

Aini Ameera Muhammad Amri S, Adawiyah Mohd R (2022) A review on islamic environmental ethics: a solution microplastics-based water. Jurnal Islam Dan Masyarakat Kontemporari 23(2):39–53. https://doi.org/10.37231/jimk.2022.23.2.233

Al-Jayyousi O, Tok E, Saniff SM, Wan Hasan WN, Janahi NA, Yesuf AJ (2022) Re-thinking sustainable development within islamic worldviews: a systematic literature review. Sustainability (Switzerland) 14(12). https://doi.org/10.3390/su14127300

Ali B, Gul P (2018) An Islamic approach towards environmental education. J Law Soc XLIX(73):63–76

Ashtankar OM (2016) Islamic perspectives on environmental protection. Int J Appl Res 2(1):438–441. www.allresearchjournal.com

Aziz F, Mahadi N, Jihad M (2018) Linking Islamic work ethics and pro-environmental behaviour: a systematic review. J Soc Sci Res Spec Is(2):249–256. https://doi.org/10.32861/jssr.spi2.249.256

Baum SD, Owe A (2023) Artificial intelligence needs environmental ethics. Ethic Policy Environ 26(1):139–143. https://doi.org/10.1080/21550085.2022.2076538

Braun T, Dierkes P (2017) Connecting students to nature—how intensity of nature experience and student age influence the success of outdoor education programs. Environ Educ Res 23(7):937–949. https://doi.org/10.1080/13504622.2016.1214866

Bsoul L, Omer A, Kucukalic L, Archbold RH (2022) Islam's perspective on environmental sustainability: a conceptual analysis. Soc Sci 11(6). https://doi.org/10.3390/socsci11060228

Chairunnisa E (2018) Peranan World Wide Fund for Nature (WWF) dalam Upaya Konservasi Populasi Badak Jawa di Indonesia. Glob Polit Stud J 2(1):72–87. https://doi.org/10.34010/gps journal.v2i1.2012

Collado S, Rosa CD, Corraliza JA (2020) The effect of a nature-based environmental education program on children's environmental attitudes and behaviors: a randomized experiment with primary schools. Sustainability (Switzerland) 12(17). https://doi.org/10.3390/SU12176817

Deplazes-Zemp A (2023) Beyond intrinsic and instrumental: third-category value in environmental ethics and environmental policy. Ethic Policy Environ 00(00):1–23. https://doi.org/10.1080/215 50085.2023.2166341

Des Jardins JR (2013) Environmental ethics: an introduction to environmental philosophy, 5th ed. Wadsworth Cengage Learning

Dzwonkowska D (2018) Is environmental virtue ethics anthropocentric? J Agric Environ Ethics 31(6):723–738. https://doi.org/10.1007/s10806-018-9751-6

Ellingson S, Vernon A, Woodley AP (2012) The structure of religious environmentalism: movement organizations, interorganizational networks, and collective action. J Sci Stud Relig. https://doi.org/10.1111/j.1468-5906.2012.01639.x

Eterović I, Buterin T (2022) Bioethical analysis of sanitary engineering: a critical assessment of the profession at the crossroads of environmental and public health ethics. Ethic Sci Environ Polit 22:13–24. https://doi.org/10.3354/esep00199

Fien J (1995) Teaching for a sustainable world: the environmental and development education project for teacher education. Environ Educ Res 1(1):21–33. https://doi.org/10.1080/135046 2950010102

Gueye MK, Mohamed N (2022) An Islamic perspective on ecology and sustainability. IntechOpen. https://doi.org/10.5772/intechopen.105032

Gulzar A, Islam T, Khan MA, Haq SM (2021) Environmental ethics towards sustainable development in Islamic perspective. Ethnobotany Res Appl 22. https://doi.org/10.32859/ERA.22.39. 1-10

Hadayani T, Utami N, Abdullah K (2021) Implementation of religious character education in online learning for elementary school. In: Proceedings of the 1st annual international conference on natural and social science education (ICNSSE 2020), 547(Icnsse 2020), pp 66–72. https://doi. org/10.2991/assehr.k.210430.010

Hayat I, Malik MS, Ali MW, Husnain M, Sharif M, Haleem A (2023) The role of Islamic environmental ethics in the alleviation of climate challenges and the preservation of ecosystem. Russ Law J XI(11s):395–404. https://doi.org/10.52783/rlj.v11i11s.1967

Hummel D, Daassa M (2019) "Walking gently on the earth": an interview with nana firman on Islamic environmental ethics. J Islamic Faith Pract 2(2):24–39. https://doi.org/10.18060/23273

Irawan B, Nasution IFA, Coleman H (2021) Applying Ibn ʿArabī's Concept of Tajallī: A Sufi Approach to Environmental Ethics. Teosofia: Indonesian J Islamic Mysticism 10(1):21–36. https://doi.org/10.21580/tos.v10i1.7204

Islam MS, Karia N, Taib FM, Ara H, Moeinzadeh S (2021) Ethico-religious green supply chain management (GSCM): embedding Islamic ethics' codes for improving environmental concerns. J Islamic Account Bus Res 13(1):157–176. https://doi.org/10.1108/JIABR-02-2021-0052

Ives CD, Buys C, Ogunbode C, Palmer M, Rose A, Valerio R (2023) Activating faith: pro-environmental responses to a Christian text on sustainability. Sustain Sci 18(2):877–890. https:// doi.org/10.1007/s11625-022-01197-w

Jackson BL (2019) Ethical principles in the modern world. Logos Univ Mental Educ Novelty Soc Sci 8(2):63–75. https://doi.org/10.18662/lumenss/25

Kar S, Tripathy M (2022) Role of religion in environmental sustainability: an Indian perspective. Problemy Ekorozwoju 17(1):96–100. https://doi.org/10.35784/pe.2022.1.09

Kontio U (2020) Environmental motives in the Buddhist ecology: A Study of Thich Nhat Hanh's ecology, engaged practice and environmental activism. Uppsala Universitet

Körtner UHJ (2022) Christian faith in creation for environmental ethics and climate protection. Khazanah Theologia 4(2):135–146. https://doi.org/10.15575/kt.v4i2.19991

Krettenauer T (2017) Pro-environmental behavior and adolescent moral development. J Res Adolesc 27(3):581–593. https://doi.org/10.1111/jora.12300

Lim HL (2019) Environmental revolution in contemporary buddhism: the interbeing of individual and collective consciousness in ecology. Religions 10(2):1–14. https://doi.org/10.3390/rel100 20120

Lowe BS, Jacobson SK, Israel GD, Peterson AL (2023) Association of religious end time beliefs with attitudes toward climate change and biodiversity loss. Sustainability (Switzerland) 15(11). https://doi.org/10.3390/su15119071

Makoul Z (2022) Environmental ethics and environmental law: a virtuous circle. Columbia J Environ Law 47(1):68–100. https://doi.org/10.52214/cjel.v47i1.9131

Mangka A, Husma A, Mangka J (2022) Pelestarian Lingkungan Hidup dalam Pandangan Syariat Islam. BUSTANUL FUQAHA: Jurnal Bidang Hukum Islam 3(2):205–221. https://doi.org/10. 36701/bustanul.v3i2.613

Manoiu V-M, Düzgüneş E, Azzeddine M, Manoiu V-S (2016) A qualitative exploration of the Holy Quran's environmental teachings. IJAEDU Int E-J Adv Educ 2(5):209. https://doi.org/10.18768/ ijaedu.43510

Marcus ZJ, McCullough ME (2021) Does religion make people more self-controlled? A review of research from the lab and life. Curr Opin Psychol 40:167–170. https://doi.org/10.1016/j.copsyc. 2020.12.001

Mauluah L, Marsigit, Wangid MN (2021) Rancangan Lembar Kerja Matematika SD/MI Berbasis Islamic Environmental Ethics. Mualimuna: Jurnal Madrasah Ibtiadiyah 6(2):70–82

Michelsen G, Fischer D (2017) Sustainability and education 1. In: von Hauff M, Kuhnke C (eds) Sustainable development policy A European perspective. Routledge. https://doi.org/10.4324/ 9781315269177

Moher D, Liberati A, Tetzlaff J, Altman DG, Altman D, Antes G, Atkins D, Barbour V, Barrowman N, Berlin JA, Clark J, Clarke M, Cook D, D'Amico R, Deeks JJ, Devereaux PJ, Dickersin

K, Egger M, Ernst E et al (2009) Preferred reporting items for systematic reviews and meta-analyses: the PRISMA statement. PLoS Med 6(7). https://doi.org/10.1371/journal.pmed.100 0097

Moleong L (2006) Metodologi penelitian. Kualitalif Sasial

Mustolikh M, Budimansyah D, Darsiharjo D, Nurdin ES (2022) Bencana Alam dan Etika Lingkungan Hidup dalam Al-Qur'an. Proc Ser Soc Sci Human 6(0281):170–176. https://doi.org/10.30595/pssh.v6i.459

Najib M, Saefullah U, Haryanti E, Haetami E, Maryati DS (2020) Environmental conservation paradigm in Islamic ecological perspective. Int J Psychosoc Rehabil 24(4):5440–5447. https://doi.org/10.37200/IJPR/V24I4/PR201640

Omar SM, Chowdhury AJK, Hashi AA (2018) Islamic ethics of waste management towards sustainable environmental health. Int Med J Malays 17(Specialissue1):193–197

Otto S, Evans GW, Moon MJ, Kaiser FG (2019) The development of children's environmental attitude and behavior. Glob Environ Change 58(August 2018):101947. https://doi.org/10.1016/j.gloenvcha.2019.101947

Page MJ, McKenzie JE, Bossuyt P, Boutron I, Hoffmann TC, Mulrow CD, Shamseer L, Tetzlaff JM, Akl E, Brennan SE, Chou R, Glanville J, Grimshaw JM, Hróbjartsson A, Lalu MM, Li T, Loder EW, Mayo-Wilson E, McDonald S et al (2021) The Prisma 2020 statement: an updated guideline for reporting systematic reviews. BMC 10(89):1–11. https://doi.org/10.21860/medflu m2021_264903

Palmer C, McShane K, Sandler R (2014) Environmental ethics. Annu Rev Environ Resour 39:419–442. https://doi.org/10.1146/annurev-environ-121112-094434

Pasaribu AG, Sipahutar RCHP, Hutabarat EH (2022) Imago Dei and ecology: rereading genesis 1:26–28 from the perspective of Toba Batak in the ecological struggle in Tapanuli, Indonesia. Verbum Et Ecclesia 43(1):26–28. https://doi.org/10.4102/ve.v43i1.2620

Perry A, Hammond N (2002) Systematic reviews: the experiences of a PhD student. Psychol Learn Teach 2(1):32–35. https://doi.org/10.2304/plat.2002.2.1.32

Petticrew M, Roberts H (2008) Systematic reviews in the social sciences: a practical guide. In: Systematic reviews in the social sciences: a practical guide. https://doi.org/10.1002/978047075 4887

Purwanto R, Istiani M, Hilda H, Marazi H, Nudin B (2022) Islamic view towards environment preservation. KnE Soc Sci 2022:11–15. https://doi.org/10.18502/kss.v7i10.11336

Ragozina SA (2023) Environmentalism in modern Islamic philosophy. RUDN J Philos 27(2):233–250. https://doi.org/10.22363/2313-2302-2023-27-2-233-250

Sadowski RF, Ayvaz Z (2023) Biblical and Quranic argumentation for sustainable behaviors toward nature. Problemy Ekorozwoju 18(1):152–163. https://doi.org/10.35784/pe.2023.1.15

Sahih al-Bukhari 2320, Book 41, Hadith 1. (n.d.). Retrieved July 11, 2023, from https://sunnah.com/bukhari:2320

Sayem MA (2019) The eco-philosophy of Seyyed Hossein Nasr: spiritual crisis and environmental the eco-philosophy of Seyyed Hossein Nasr: spiritual crisis and environmental degradation. Islam Stud 58(2):271–295

Sen S (2021) The eco-theology of the bhagavad gītā: a multi-layered ethical theory. Religions 12(4). https://doi.org/10.3390/rel12040241

Siddique A (2024). Behavioral consequences of religious schooling. J Dev Econ 167(27331). https://doi.org/10.1016/j.jdeveco.2023.103237

Siombo MR, Adi EAW (2021) Desa Berwawasan Lingkungan Melalui Sinkronisasi Kewenangan Desa dan Pelibatan Masyarakat dalam Proses Persetujuan Lingkungan. Bina Hukum Lingkungan 5(3). https://doi.org/10.24970/bhl.v5i3.218

Stephens DW (2019) From environmental stewardship to environmental holiness: the evolution of methodist environmental witness, with a focus on climate change. J Relig Ethic 47(3):470–500. https://doi.org/10.1111/jore.12281

Strubell E, Ganesh A, McCallum A (2020) Energy and policy considerations for deep learning in NLP. ACL 2019—57th annual meeting of the association for computational linguistics. In: Proceedings of the conference, vol 1, pp 3645–3650

Sukino S, Utami SP (2020) Islamic Religious education models in preventing negative behaviors of youth and adolescents. Tarbawi: Jurnal Keilmuan Manajemen Pendidikan 6(02):193. https://doi.org/10.32678/tarbawi.v6i02.3539

Sunan Ibn Majah 425, Book 1, Hadith 159. (n.d.). Retrieved July 11, 2023, from https://sunnah.com/ibnmajah:424

Sutarto S (2023) Internalization of Islamic educational values on clean living as an effort for the formation of environmental care attitudes for elementary school students. Jurnal Konseling Dan Pendidikan 10(3):555–563. https://doi.org/10.29210/181600

Syafi'i I, Mardiyah M (2023) Implementasi Kegiatan Keagamaan dalam Meningkatkan Kecerdasan Spiritual Siswa. Jurnal Keislaman 6(1):256–267. https://doi.org/10.54298/jk.v6i1.3715

Tanzeh A, Junaris I (2019) Principal policy in developing Islamic student character. Khatulistiwa: J Islamic Stud Inst Res Commun Serv 9(1):5–14. https://doi.org/10.24260/khatulistiwa.v9i1.1299

Twardziłowski T (2020) The command to rule over the creation (Gen 1:26–28) in the ecological hermeneutics of the bible. Collectanea Theologica 90(5):9–32

Wahab MA, Ismail Y (2019) Mas'uliyya and Ihsan as high-performance work values in Islam. Int J Econ Manage Account 27(1):187–212

Wals AEJ, Benavot A (2017) Can we meet the sustainability challenges? The role of education and lifelong learning. Eur J Educ 52(4):404–413. https://doi.org/10.1111/ejed.12250

Watsiqotul S, Agung L (2018) Peran Manusia Sebagai Khalifah Allah di Muka Bumi Perspektif Ekologis dalam Ajaran Islam. Jurnal Penelitian 12(2):355–378. https://journal.iainkudus.ac.id/index.php/jurnalPenelitian/article/view/3523/pdf

Yulianingsih U (2023) Pembiasaan Pagi Sejak Madrasah Dalam Menanamkan Perilaku Religius. Fashluna 4(2):119–130. https://doi.org/10.47625/fashluna.v4i2.511

Yusmaliana D, Fitriana F, Zakaria GAN, Kurbiyanto A (2023) Ecological education in Islamic religious learning based on creative imagination. In: Sharifi A, Simangan D, Kaneko S (eds) Bridging peace and sustainability amidst global transformations. Springer, Singapore. https://doi.org/10.1007/978-981-99-7572-3_4

# Southeast Asia's Sea Nomads: A Case for Decolonizing Environmental Peacebuilding Scholarship by Integrating Indigenous Sama-Bajau Perspectives

Mikaela Francine D. Lagarde

**Abstract** This chapter adopts a critical and context-specific approach to expanding environmental peacebuilding theory and practice. It proposes a decolonizing approach to environmental peacebuilding, where peace and sustainability are pursued beyond mainstream, Western-centric paradigms, and instead are informed by Indigenous practices, traditional ecological knowledge, and local wisdom that harmoniously integrate humans and nature. Centering Indigenous epistemologies and ontologies encourages a rethinking and reshaping of perspectives and actions toward peace, the environment, and Indigenous knowledge. Southeast Asia's Sama-Bajau (Sama, Sama Dilaut/Pala'u, Bajau Laut, Badjao, Bajo, or the "sea nomads") offer an illustrative case. These Indigenous maritime communities often face marginalization and neglect due to their unique lifestyle and dependence on the sea, as well as increased vulnerabilities due to climate change and environmental degradation. In essence, decolonial peace for the Sama-Bajau goes beyond the absence of conflict; it involves acknowledging social injustices, embracing cultural perspectives, and integrating environmental sustainability. Environmental peacebuilding becomes a way forward by fostering a holistic approach that complements the sea tribes' agency and ability to shape their own path toward durable peace.

**Keywords** Environmental peacebuilding · Decolonization · Indigenous Sama-Bajau maritime communities

## 1 Environmental Peacebuilding

Environmental peacebuilding is a process that integrates peacebuilding with environmental protection and/or natural resource management in order to address violent conflicts and build community resilience (Dresse et al. 2019; Ide et al. 2021a). It theorizes that, at a minimum, low-level cooperation on shared environmental concerns

M. F. D. Lagarde (✉)
School of Law, Queensland University of Technology (QUT), Brisbane, Australia
e-mail: m.lagarde@hdr.qut.edu.au

© The Author(s), under exclusive license to Springer Nature Singapore Pte Ltd. 2024
A. Sharifi et al. (eds.), *Navigating Peace and Sustainability in an Increasingly Complex World*, World Sustainability Series, https://doi.org/10.1007/978-981-97-8772-2_3

can contribute to dialogue and trust building between divided groups (Bruch et al. 2021), thereby reducing tensions, regardless of the cause of conflict. Such predominantly technical and depoliticized cooperation is expected to "spillover" into wider forms of political cooperation (Carius 2006). Some view environmental peacebuilding as a subset of a broader approach, environmental peacemaking (Conca and Dabelko 2002; Harris and Mische 2008), but this chapter considers these terms interchangeable.

Many Indigenous cultures and non-Western societies have long practiced principles akin to environmental peacebuilding, albeit in a more holistic sense. For example, the Navajo Nation or Diné have spiritual beliefs and practices that emphasize "*hózhó*"—the concept of beauty, harmony, balance, stability, peaceful energy, well-being, and a sacred perfect rhythm—considering humans, nature, and the cosmos as one and interrelated (Kahn-John and Koithan 2015; Neezzhoni 2010). This belief maintains environmental stewardship and community harmony, which align with modern environmental peacebuilding goals but do not quite fit the mainstream (Western) paradigms.

Many of the theoretical frameworks and practices informing mainstream environmental peacebuilding have a Western-liberal core, including liberal peace (Richmond and Mac Ginty 2015), market-based conservation policies (Collins et al. 2021), and natural resource management (Searle and Muller 2019), resulting in the influence of Western values and solutions on diverse cultural contexts. Moreover, top-down environmental peacebuilding projects often prioritize Western methods and technologies, marginalizing Indigenous knowledge. For instance, national conservation strategies like no-take marine protected areas clash with a maritime-oriented tribe's food sovereignty and traditional, spiritual, and livelihood practices. Such attempts at spatial and temporal exclusion prioritize biological/scientific criteria over local community needs, undermining food security (Stacey et al. 2017) and cultural preservation (Do Khac 2018).

Decolonizing environmental peacebuilding theory and practice involves resolving conflicts in a way that liberates and dismantles colonial influences and questions the Cartesian dualistic worldview. This process entails reversing the effects of colonial perspectives, institutions, and their consequences (Schirch 2022). For there to be meaningful change, environmental peacebuilding needs to consider heterogeneous, "extra-local" peacebuilding networks of Indigenous groups and traditional communities, diverse temporalities, and pluriversal alternatives, as well as recognize environmental peacebuilding's link to colonialities of power. Doing so averts the violence against Indigenous people, their culture, and the ecosystems.

Balancing the views, I take the position that environmental peacebuilding is not solely a colonial or Western/Euro-centric concept; rather, it represents a dynamic approach that, when implemented thoughtfully, can integrate diverse perspectives and practices. This chapter advocates for embedding a decolonial perspective in environmental peacebuilding theory and practice, using the Sama-Bajau as an illustrative case.

## 1.1    Expanding Environmental Peacebuilding

Environmental peacebuilding is relevant in an armed conflict situation, often as a starting point—a means for conflicting parties to build mutual trust through environmental cooperation. I argue that the elements of an armed conflict and conflicting parties in existing literature limit the potential for environmental peacebuilding theory and practice. I propose expanding the conceptual framework to also include situations where there is no armed conflict or threat thereof, and the parties are not fighting in a traditional sense, but there is nevertheless enduring structural violence and oppression. Such a proactive approach coheres with Galtung's concept of positive peace (Galtung 1969)—where communities not only live without conflict but also experience justice, socio-economic well-being, and a sustainable relationship with their environment.

The Westphalian State is considered the conflicting party as it upholds colonialist-thinking that devalues Indigenous, sea-based nomadic lifestyles. The ontology of the modern state system disregards and erases Indigenous life-worlds (Simon and Mona 2023). Indigenous peoples regard the sea in different ways, including as a key political entity—a place "on which claims are made, to which meanings are attached, and over which political conflicts erupt" (Do Khac 2018, p. 37)—an "animated seascape" (McNiven 2004) inhabited by humans, nonhuman beings, and spirits controlling the waters (Andaya 2017). This cannot easily be explained by Western ontologies. Even critical geopolitics, which accounts for the social construction of place (Dodds et al. 2013; Tuathail and Toal 1996), appears limited in explaining "more-than-human" entities and their interactions with power and space (Koopman et al. 2021). The Sama-Bajau perspective is multi-spatial, in which the sea is a shared cultural space and "place" is constructed depending on the meanings derived from their experiences with their natural surroundings and social interactions (Lowe 2003), including with the local sea spirits. The Sama-Bajau are essentially terraqueous global–local communities, concepts that challenge aspects of mainstream interstate (state-centric) and/or intrastate (local) environmental peacebuilding frameworks unless a decolonial perspective is integrated. For the Sama-Bajau to achieve positive peace, one can explore environmental peacebuilding as a starting point, albeit from the critical lens of Indigenous and decolonial thought.

## 2    Sama-Bajau Maritime Communities

Before I begin this section, I must preface that I do not have any special right to represent or speak for the Sama-Bajau; I am but an ally. My interest in aquatic communities and decolonial epistemology roots from my belongingness to ethnolinguistic groups native to the river (Kapampangans of Central Luzon, whose name is derived from "pampang" or riverbank) and the Visayan coast (Negrenses of Negros Occidental, whose first inhabitants were the Aboriginal Ata) in the Philippines. This chapter also does not engage in a rhetoric of emancipation or "salvation" of the Sama-Bajau; rather, it simply aims to foreground how environmental peacebuilding may be complicit in marginalizing, if not erasing, certain life-worlds and knowledge systems unless the disciplinary assumptions of the field are critically engaged with. Indigenous perspectives are often constrained by stereotypes and historical categorizations, preventing their exploration beyond Western-imposed ideological boundaries (Hunt 2014). This chapter's analysis focuses on the Sama-Bajau of central maritime Southeast Asia, excluding other similar sea-oriented communities,[1] and building on the work of many scholars who have conducted the research on the ground.

Southeast Asia is the world's largest archipelago and the traditional home to the aquatic Indigenous communities of the Sama-Bajau. Extensively dispersed across maritime zones, they are commonly found in three territorial areas: the Mindanao island group in southern Philippines; Sabah in Malaysian Borneo (a territory under Malaysia's control but disputed by the Philippines); and eastern Indonesia. The Sama-Bajau rely on the sea for their cultural identity, food, livelihood, way of life, and social organization. Famous for their free diving abilities, navigational expertise, fishing techniques, and deep knowledge of the seascape, they have been highly valued in transportation and trade by land-based polities. Their mobility historically shaped social relationships, economies, and cultural dynamics among different communities, facilitating the exchange of goods, ideas, culture, technologies, and regional trade networks from Southeast Asia to the Indian Ocean regions, China, and Australia (Bellina et al. 2021; Hoogervorst 2012; Maglana 2016; Stacey 2007).

While many writers categorize them collectively as "sea people" or "sea nomads," Sama-Bajau communities have significant internal diversity and fragmentation; boat types, languages, religious practices, and fishing techniques vary among different groups (Nimmo 1968). To attain a nuanced understanding, this chapter uses the ethnonym "Sama-Bajau" to refer to the maritime communities dependent on the sea as a source of life and livelihood, whether currently nomadic (boat-dwellers at sea), semi-nomadic (living in stilt houses on reefs in non-urban coastal zones), or those who have recently abandoned such lifestyles due to external pressures such as forced displacement.

## 2.1   Legacies of Colonialism

The Sama-Bajau communities have faced discrimination and marginalization due to their distinctive cultural practices and beliefs. Their history has been marked by violence, dispossession, alienation, exile, and cultural erasure as ongoing impacts of colonization that began in the late fifteenth century. During the colonial period, it was the British in Sabah; the Dutch in eastern Indonesia; and the Spanish, American, and

---

[1] Other sea tribes in the region, such as the Mokken/Moklen of the Mergui Archipelago in Myanmar and Thailand and the Orang Laut/Urak Lawoi of the Thai-Malay Peninsula and Riau Archipelago in Indonesia, are not covered by this chapter.

then Japanese in parts of Mindanao. Occupying powers described them in pejorative terms: e.g., a "wandering irresponsible sort of race, rather low down in the scale of humanity" (The British North Borneo Herald (1886), cited in Marshall et al. 2019). They were also called "sea gypsies", "gypsy" being a known racial slur for the nomadic Roma in Europe with whom the Spanish and British colonizers compared the Sama-Bajau (Sopher 1965; Stacey 2007).

The problem of external colonialism transitioned to internal colonialism in the form of the modern state system, which introduced borders in an otherwise border-less sea. The concept of internal colonialism underscores the same unequal power dynamics and exploitation within national borders—where a dominant group (i.e., Westphalian state, represented by a post-colonial government) has occupied the role previously played by the colonizers and repeated the same socio-economic exploitation and domination upon the natives (González Casanova 1965). Colonial-thinking draws life from Eurocentric patterns of hierarchical dualisms—e.g., white/non-white, settler/nomad, man/woman, modern/backward, and human/nature—where the first of the pair is perceived as superior to the second (Elvey 2020; Lowe 2003). The enduring structures of colonialism have shaped the political (e.g., citizenship), legal (e.g., native title), social (e.g., racism), economic (e.g., global neoliberal system of capital and labor), and cultural (e.g., Western epistemic privilege) systems and institutions today.

## 3 Different Forms of Violence

Peace research and environmental studies delineate various forms of violence, including cultural, psychological, structural, epistemic, and ecological dimensions, each having significant human and environmental costs. For the Sama-Bajau, structural violence is evident through laws, formal institutions, and societal structures defined by class, hierarchy, and race. These factors often leave them stateless or treated as second-class citizens. State actions, influenced by historical patterns of colonial oppression, marginalize these communities due to their nomadic lifestyle, Indigenous identities, and/or statelessness.

### 3.1 Hard Borders as Structural Violence

The highly mobile Sama-Bajau historically did not have permanent land-based territoriality (Clifton 2014); their Indigenous marinescape, which this chapter refers to as "maritory," overlaps the present-day maritime zones of three states. The Sama-Bajau's deep connection to their maritory is rooted in constant movement, with flexible social structures and reliance on fishing, harvesting of marine resources, and other sea-going activities (Bellina et al. 2021). Unfortunately, their mobility across the vast expanse of seas has been limited by post-colonial governments (Clifton

2014). The legacies of colonialism resulted in the founding of newly independent nation-states with hard borders that have "immobilized" or restricted their movements (Barrena et al. 2022). Examples of such restrictions include those set by the UN Convention on the Law of the Seas (UNCLOS), national laws demarcating territory, and administrative rules on no-fishing zones. Often disregarding Indigenous maritory, such demarcation of national boundaries illegalizes, if not penalizes, their sea-borne nomadic lifestyle and cultural practices—their multi-spatiality—pushing them into "unlawful" migration patterns (Roxas-Lim 2017).

Under UNCLOS, it is the state, not the maritime-oriented Indigenous people, that has sovereignty over the land territory, internal waters, and territorial sea; for archipelagic states like the Philippines and Indonesia, this sovereignty extends to their archipelagic waters (Part II, Article 2[1]). Even in legal systems heavily influenced by common law, such as Australia, where sea country is acknowledged as traditionally owned and managed by Indigenous peoples, native title claims over aquatic territories have found limited success (Butterly and Richardson 2016). As a result, the Sama-Bajau are dependent on the beneficence of the state rather than any ancestral rights over the sea. Lacking legal recognition and protection as sea-based First Nations, they face issues of displacement and involuntary adoption of a sedentary lifestyle.

## 3.2   Statelessness

Although the transnational Indigenous sea tribes predated the creation of modern states and have existed outside their realm for millennia (Andal 2023), the founding of the Westphalian state rendered many of them stateless. Revolving their lives around the sea means most are poor, not formally schooled, seldom read or write, and have no legal nationality documents (Allerton 2014; Nor 2017). Without legal documentation, they cannot access citizenship rights, such as government services and rights to stay and fish. Being stateless leaves them highly vulnerable to abuse and exploitation. In Malaysia, the undocumented Sama-Bajau, including those born in Sabah who have never left the island, are suspected by the government to be "illegal" Filipino immigrants and a threat to Malaysian sovereignty (Acciaioli et al. 2017; Hoogervorst 2012; Ziegler 2022). Several Sama-Bajau communities in Sabah have been arrested, detained, and deported to the Philippines or time-restricted from entering their ancestral waters (Chiew 2019; Cooke and Acciaioli 2021; Marshall et al. 2019). Indonesian security forces would likewise deport undocumented sea tribes for being "citizens of a foreign country" (Stacey et al. 2017; Roxas-Lim 2017). In the Philippines, Sama-Bajau tribes affected by the armed conflicts in Mindanao are internally displaced or would cross maritime borders, particularly to Sabah, as refugees, where they risk being sent back to the Philippines (Aoyama 2012; Medina 2016). Fishing close to the international maritime borders also risks their capture for illegal fishing as they lack the technology to warn them of "unlawful" border crossing while at sea (Stacey 2007). Statelessness doubly victimizes Indigenous people through marginalization, discrimination, loss of rights to fish in their maritory, loss of cultural identity, limited access to government services, and lack of legal protections. Currently, the Philippines is the only Southeast Asian country with a National Action Plan to end statelessness (UNHCR Philippines, 2019), i.e., by providing birth certificates to at-risk Sama-Bajau in the country (UNICEF Philippines, 2024). There are about one million Sama-Bajau who are "citizens of nowhere" (Ziegler 2022).

## 3.3  Lack of Maritory Rights

Not having maritory rights makes them one of the most marginalized Indigenous communities in the world. For instance, Republic Act No. 8371 or Indigenous People's Rights Act (1997) recognizes that land-based Indigenous Cultural Communities/Indigenous Peoples (ICC/IPs) in the Philippines have native title to their pre-conquest ancestral lands as well as access to resources within their ancestral domains (i.e., lands, inland waters, and coastal areas in the Philippines, excluding the wider marine space). However, the legal regime applicable to native sea title for maritime-oriented, nomadic ICC/IPs is questionable, unless the concept of the sea is deconstructed to mean Indigenous "land" as well (Somiah 2022). While the UN Declaration on the Rights of Indigenous Peoples (2007) and other declarations involving ICC/IPs rights and interests carry some political leverage, these UN resolutions are non-binding and lack specific guidance for local-level application (Butterly and Richardson 2016). Indigenous claims to sea and sovereignty fundamentally challenge the political, economic, and spatial hegemony of the state. The closest possibility of a firm legal right is the state's assertion of water use and access under UNCLOS.

Firm legal rights are essential for the Sama-Bajau to protect their nomadic lifestyle, access to resources, and cultural heritage. Without such, they will remain trapped in the dole-out and developmental system, if not deportation, that has stripped them of their agency. A longer-term solution is to evolve public international law to recognize the Sama-Bajau identity and culture, their maritory rights, and their status as a transnational maritime-oriented Indigenous people. Until then, one must operate within the existing legal framework, e.g., the law of the sea, customary international law, and fundamental human rights principles. Parts of UNCLOS can be used to advocate for the Sama-Bajau. UNCLOS guarantees traditional fishing rights in the archipelagic waters of a directly adjacent neighboring coastal state (Article 51[1] vis-à-vis Article 47[6]), albeit not in that state's exclusive economic zone (EEZ) ("South China Sea Arbitration," 2016). Meanwhile, customary international law arguably permits traditional fishing even in another coastal state's EEZ (North 2017). The International Bill of Human Rights also collectively establishes fundamental political, social, economic, and cultural rights and freedoms for all individuals. These come with a caveat: it is one thing to have legal protections enshrined and another to have a home state (generally) invoke these rights on behalf of the Sama-Bajau. The requirement for them to have a bond of nationality remains a sticking point. Thus, relying solely on international law for recognition of their maritory rights is insufficient, thereby limiting options.

## 3.4 Loss of Collective Indigenous Identities

The Sama-Bajau's rights to sail and fish within Malaysian, Indonesian, or Philippine internal and territorial waters hinge on obtaining citizenship from the respective states. For those who acquired citizenship, issues remain that arise from the incompatibility between Indigenous and modern state systems. Citizenship and mobility are framed from the colonial understanding of state/territoriality/legal personhood (Jegen 2023) rather than the Sama-Bajau ontologies of collective identities/multi-spatiality/relational networks (Marshall et al. 2019; Pauwelussen 2016). Indigenous peoples have always been collective with a communal identity (Berry 1988); Western epistemologies and ontologies individualized and "citizen-ized" (ciudadanizan) (Macas 2010) them, fitting their collective identity and political community membership within a statist framework. The Sama-Bajau have historically been a "world tribe" interacting and living in harmony with other sea tribes and nonhuman beings, maritime features, ancestral spirits, and inland communities (Clifton 2014; Tahara 2021).

In Malaysia, the government policy of *Bumiputera* (meaning, "children of the soil") forces the central authority-driven ethnicization and homogenization of otherwise unique and multiple Indigenous identities in Peninsular and East Malaysia, including the Sama-Bajau (Somiah 2022). The logic of Bumiputeraism is rooted in the institutionalization of a state-constructed national identity, Bumiputera—a "highly specific form of Malayness"—to identify the beneficiaries of top-down developmentalist policies (Ibrahim 2021, p. 7). Article 161(A)(6)(b) of the Malaysian Federal Constitution legally considers (documented) Sama-Bajau tribes in Sabah as Bumiputera or ethnically Malay, notwithstanding their unique indigeneity, translocality, and non-"Malayness" (Cooke and Acciaioli 2021; Hoogervorst 2012; Maglana 2016; Somiah and Santo Domingo 2021). In Indonesia, the government considers them a target for mainstream socialization (Acciaioli 2001; Lowe 2003), seeking to "develop" them culturally and economically in line with the Indonesian state philosophy of Pancasila (Clifton 2014). Severing Sama-Bajau identity, through law or policy, from that of their Indigenous community and culture removes their agency for self-identification. State assimilation allows for the centralization of mechanisms of social and environmental control (Cooke and Acciaioli 2021)—the same mechanisms that reproduce and reinforce the coloniality of power.

## 3.5 Cultural Violence

Cultural systems shaped by the coloniality of power assert that only European cultures embody true modernity, defined by traits such as capitalism, rationality, neoliberalism, and science. "White"/Western (Global North) heteropatriarchal ideals, experiences, and epistemology are normative, while those antithetical to these categories are judged and differentiated accordingly as the Other (Moreton-Robinson 2021).

These racialized and genderded systems reinforce Eurocentric norms through state and economic mechanisms (Nuraini 2016; Quijano 2000). The exclusionary "invention of the Other" (Castro-Gómez 2019) allows for colonial representations of the Sama-Bajau through racist stereotypes, such as being labeled as savages "low" on humanity. Indigenous people who fish and dive for subsistence are still viewed as civilizationally backward, with lifestyles deemed obsolete in the age of progress and commercial expansion (Lowe 2003; Stacey et al. 2017). The devaluation of their way of life and the failure to acknowledge and allocate space for their cultural preservation constitute cultural violence.

The pressure that is imposed by modern states or external forces to adopt a sedentary lifestyle has led to a decline in the traditional value and respect for the sea people's navigational skills and sea-related expertise (Andaya 2019). Marine environmental degradation and overfishing by industrial fishing fleets force some Sama-Bajau communities to settle on land (Ferguson et al. 2022; Illouz and Nuraini 2021). Settling on dry land, they are further marginalized (Medina 2016), as government policies often seek to "modernize" their lifestyles from a developmentalist lens without regard for culture or sustainability (Andal 2023). In a culture with a historically fluid sense of boundaries and sea-oriented lifestyles, their displacement, reterritorialization (forced resettlement), and cultural assimilation (via sedentarism) eliminate their autonomy and sovereign prerogative.

Bureaucrats and conservationists also tend to blame them for damaging the marine environment. As they are simply transient, the sea tribes are believed to be uninterested in preserving biodiversity (Lowe 2003), lacking the incentive to respond to declining catches with community-based fisheries management (Stacey et al. 2017). Unsustainable resource use and destructive fishing practices have been attributed to the Sama-Bajau with little direct evidence, as they are easier to blame than governmental corruption or corporations, other coastal people's activities, or industry-scale fishing (Clifton 2014). This narrative is exploited to deflect attention from the latter's own destructive practices. A study finds that the Sama-Bajau utilize safe fishing gear that does not threaten marine life (Rahmawati and Kahirun 2020). Using them as scapegoats for environmental destruction is another form of cultural violence.

Losing their sense of cultural identity, dignity of work (Simonin 2015), and self-worth also impairs Indigenous people's psychological and emotional well-being. Moreover, their relocation into urban areas often leaves them excluded from formal employment, begging for survival, and trapped in an intergenerational cycle of poverty (Chia 2016). In urban centers like Manila and Davao, the term "Badjao" is synonymous with "beggars" (Cooke and Acciaioli 2021). All these factors widen socio-economic inequalities (Roxas-Lim 2017).

## 3.6 Epistemic and Ontological Violence

Western epistemology upholds the land-sea dichotomy of thinking *about* water (separately, from a landed perspective) rather than *with* water (integrally, from an aquatic perspective) (Chen 2013). That is to say, European imperial/colonial thinking sees *terra firma* as a home and the seascape as an "other" place—a mere gateway to distant lands for conquering and/or a source of natural resources in itself. As rational land mammals, humans travel on water via ships but do not live in it (Chen et al. 2013). In contrast, the Sama-Bajau cosmos is pluriversal, involving the visible and invisible, with humans, nonhuman beings, and spirits coexisting in the physical and spiritual worlds (Bottignolo 1995; Sather 2001). They consider "community" to include both human and nonhuman, mobile and translocal entities. They share close and interdependent connections with fish and sea spirits, embodying a sense of community that transcends human/nature and social/ecological dichotomies (Pauwelussen 2016). They are independent from their neighboring inland communities while simultaneously dependent on them, with interactions and exchanges shaping the Sama-Bajau presence in the region (Bellina et al. 2021).

Moreover, in the eyes of the hegemonic Western knowledge system, naturalized Indigenous epistemologies and ontologies are considered primitive lore. The dominant "ways of knowing and thinking about nature" (Gibbs 2010) privilege Western sovereignty and the "objective" and "universal" value of science (Searle and Muller 2019). The written literature of the colonial tongue carries de facto power (Abdi 2009) over Indigenous oral histories and performative expressions. Sama-Bajau epistemes are "informed and afforded by their mobile practices, rich marine experiences, and Indigenous wisdom" (Andal 2023, p. 10) rather than by Western universalist ideals. Their folk narratives preserve and showcase their cultural practices, values, and beliefs through oral traditions. There are few archaeological records on the sea people; not much is written about them, or by them, due to their transitory lifestyles, culture that rejects the desire for material wealth, and lack of a written language (Amat and Samad 2013; Hoogervorst 2012). Second-hand sources primarily convey their stories to outsiders (Andaya 2019; Jubilado 2010; Sopher 1965). They learn how to thrive in their marine environment from their ancestors (Bellina et al. 2021). They preserve and pass on their culture, both intangible (e.g., music, dances, rituals, oral literature like tekodon, folk narratives, and a profound understanding of the natural and supernatural realms) and tangible (e.g., architecture, carving, and shipbuilding), predominantly via words, symbols, and performances (Amat and Samad 2013; Jubilado 2010; Nuraini 2016).

A positivist-rationalist lens—adopted from colonial structures and epistemes (Ide 2019)—restricts its applicability to the Sama-Bajau whose life-worlds cannot be explained by purportedly objective and scientific truths. For instance, low fish catch is attributed to the spirits' control (Stacey et al. 2017). Eroding such traditional belief-driven fishing practices by forcing the adoption of a Western lens simply heightens their vulnerability as a fishing community (Simonin 2015). Moreover, post-colonial neoliberal extractivist policies have led to a hegemonic order that violently suppresses alternative approaches to valuing, understanding, and existing. The marginalization of the Sama-Bajau ancestral knowledge, accompanied by the disregard for their voices and perspectives, results in epistemic and ontological violence. To address this intellectual colonization, one can draw lessons from Indigenous and decolonial theorizing.

### 3.7 Ecological Violence

Tensions between a colonial order, currently embodied in the Westphalian state system, and Indigenous people have long existed, but the rise of capitalism has increased their intensity. The Westphalian state has served, if not promoted, capitalist goals. The neoliberal global economy prioritizes large-scale fishing operations over traditional or artisanal fishing (Cooke and Acciaioli 2021). Development models based on unsustainable extraction, such as industrial fishing and offshore gas/oil exploration, heavily rely on the colonial-capitalist thinking of accumulating by dispossessing. This approach allows those in power to seize and exploit nature and human bodies for wealth generation that propels "development" (Cusicanqui 2012; Mezzadra and Neilson 2017; Teschke and Lacher 2007). Ecological violence manifests in the degradation of the seas caused by the neoliberal development model built around profits and commercial claims to marine resources rather than on collective well-being (Karmakar and Chetty 2023). The forced migration from sea to land of some Sama-Bajau communities hinders the maintenance of environmentally sensitive practices integral to their sea-oriented lifestyle, thereby exacerbating the ecological violence inherent in Western consumerist capitalism.

### 3.8 Climate Change as Violence

Climate-related burdens exacerbate the already entrenched harms brought upon Indigenous people by the drivers of anthropogenic environmental degradation (Whyte 2019). Warmer oceans significantly threaten global food security and human and ecosystem health, leading to biodiversity loss and increasing the risks of severe storms. Coping with rising sea levels, unpredictable storms, and other extreme threats requires adaptation to diverse and challenging conditions. Climate change disproportionately affects Indigenous peoples and introduces more uncertainty to their future (Elvey 2020). It impacts the sea tribe's traditional livelihoods, particularly fishing and seaweed farming (Stacey et al. 2017). Amidst this global ecological crisis, Sama-Bajau tribes may need to find alternative livelihoods (Rahmawati and Kahirun 2020).

## 4 Need for Decoloniality

Decolonization considers the subaltern—the "voices from below"—where oppression is a lived experience. Instead of applying universalist principles, it scrutinizes the ways in which local and international peacebuilding actors can support local and community mobilization as well as promote contextualized, culturally appropriate, and decolonized peace. Decolonization reclaims not only the physical space but also the epistemology and ontology of the communities inhabiting it, thus undoing the "blinders" that "render invisible the vast majority of today's planetary (and extra-planetary) organisms and beings" (Watson 2014, p. 77). Decolonial peace—"peace across spaces" (Stavrevska et al. 2022)—meaningfully dismantles the coloniality of power and the Cartesian dualistic worldview constituted within it. This is achieved by reviving customs and traditions, recovering capacities for promoting the flourishing of life, and establishing new forms of harmonious coexistence (Grim 2014; Valencia and Courtheyn 2023). Decolonial peace entails conceptualizing peace relationally—i.e., by improving human-Earth relations—and adopting decoloniality to undo oppressive systems that cause harm.

Given that Sama-Bajau concerns transcend three Southeast Asian states and that the Westphalian system deals with the state as the principal actor, is there a regional cooperative scheme or process available for addressing the various forms of violence they face? At this juncture, I introduce the potential for decolonized environmental peacebuilding to address some of these concerns. At a local level, there are elements of grassroots environmental peacebuilding, such as conflict prevention, community empowerment, and inclusive governance, that can be adopted and expanded. At an interstate level, decolonizing environmental peacebuilding is a good starting point for Indonesia, Malaysia, and the Philippines to begin cooperating on Indigenous-driven marine protected areas by leveraging the traditional ecological wisdom and ancestral knowledge of the sea tribes. This not only addresses environmental concerns but also empowers these communities, contributing to their agency and resilience. Consistent with environmental peacebuilding theory, the collaborative interactions among the Sama-Bajau and the states involved can spillover into broader joint problem-solving efforts, such as addressing legal and political matters, including ending statelessness and recognizing Sama-Bajau maritory rights.

Environmental peacebuilding is not the singular solution to all types of violence. It primarily focuses on addressing environmental stressors related to natural resources, as well as promoting sustainable environmental practices and community resilience, to bring forth positive peace. While environmental peacebuilding plays a crucial role in trust building between conflicting parties, achieving holistic peace involves addressing multiple dimensions of the conflict. Environmental peacebuilding intersects with social, economic, ecological, and political aspects, and its effectiveness is enhanced when integrated with broader peacebuilding strategies that tackle root causes, human rights issues, and social justice concerns. The sustainability of peace and the natural environment is a key consideration in environmental peacebuilding. By promoting environmental sustainability and equitable resource management, it contributes to long-term stability and resilience against future conflicts.

# 5   Decolonizing Environmental Peacebuilding

This section lists ways to pursue environmental peacebuilding through a decolonized lens to help build positive peace for the marginalized sea-dwelling communities and center a decolonial perspective in environmental peacebuilding scholarship. Without critical reflection, the topic attracts ethical dilemmas requiring delicate navigation; for instance, my positionality as a non-Sama-Bajau woman from the Philippines while doing research in a settler-colonial land in Australia requires a heightened awareness of my role, biases, and impact on the Sama-Bajau communities. Nevertheless, I remain optimistic that these challenges can be overcome.

To begin with, my chapter recommends decolonizing the methodological approach, in addition to decolonizing theory and practice, as an avenue for future research. This involves establishing ethical and culturally sensitive partnerships and collaborations with the Sama-Bajau, as co-producers of Indigenous knowledge, to understand and (re)construct a more comprehensive history as well as systematically analyze their unique identities, lived experiences, oral traditions, belief systems, and life-worlds. I am mindful that even as my chapter builds on the labor of many Indigenous scholars, decolonial experts, and researchers on the ground, the absence of a Sama-Bajau's direct participation in my work limits its decolonizing potential. Decolonizing methodology involves shifting from "looking at" (implicitly accepting hegemonic ideology) to "listening to" (understanding the information based on others' lived experiences) Indigenous ideas and information, as well as recognizing the asymmetry of power between settled land-based communities and mobile sea-oriented ones.

The oral history of marginalized voices must be understood, documented, and shared in a meaningful and decolonized way—by training for the skills needed to carry their stories and history in cultural context (Srigley and Sutherland 2018). This is preferred to Euro-Western theory and academic practice that simply provide parallel discourses to Indigenous contexts (Todd 2016) or co-opt Indigenous and traditional approaches (Mac Ginty 2008). Indeed, Indigenous people can hold their meaning rather than have the meaning reconstructed by Western ontology. John Borrows (2016), an Indigenous (Anishinaabeg) writer, advocates teaching the law of Indigenous peoples in line with Indigenous frameworks and culturally appropriate ways. Indigenous knowledge systems, rooted in customary law that Ancestral people follow, can be used for durable peace. Having local people as co-authors of their own peace promotes local ownership, cultural relevance, and community empowerment. Durable peace is realized where Indigenous knowledge systems are considered.

Indigenous modeling of life has more to offer than what Western-centric approaches assume. Indigenous Kwakwaka'wakw scholar Sarah Hunt suggests turning to Indigenous ways of understanding the world beyond the limits of what Western academic study accepts. One can learn from "Indigenous knowledge and the work of Indigenous thinkers (scholars, elders, community leaders, activists, community members)" for "place-specific practices" (Hunt 2014, p. 27). More specifically, there are ways to achieve this that are particular to the circumstances of the Sama-Bajau:

First, decolonize notions of time and place-based conceptualization of futures. For the Sama-Bajau, time is cyclical, fully looping back on itself, rather than linear. As the time perspective is founded on spiritual beliefs, such as social rhythms, lived experiences, spiritual understandings of hope and fear, cyclical dynamics, as well as feedback between the world and the individual, discussing the "future" becomes complex (Clifton and Majors 2012; Simonin 2015). Decolonial grassroots environmental peacebuilding must be deeply intertwined with local spiritual and communal ties. Ignoring this intertwining potentially overlooks the benefits of traditional or Indigenous ecological knowledge and practices. For the Sama-Bajau, sea spirits may be unpredictable, but they respond to pleas to ensure the success of sea-going activities (Andaya 2017). On the reverse side, the Sama-Bajau belief that time unfolds as a series of unrelated events—where actions today impact only the present and where future conditions depend solely on future deeds—can undermine the ecological rationale of long-term fisheries management (Clifton 2014). Both Indigenous cosmology and science have a role to play; these are not mutually exclusive. Considering not only ecological data, as is common in (western) conservation practices, but also the Indigenous community's deep cultural understanding of nature is a holistic approach that honors both scientific insights and Indigenous wisdom. Understanding Indigenous beliefs in sea spirits is crucial for environmental peacebuilding as it informs culturally sensitive approaches, respects Indigenous communities' sacred relationship with their environments, fosters community engagement, and aligns conservation efforts with local values, enhancing the effectiveness and sustainability of environmental initiatives.

Decolonized environmental peacebuilding envisions and constructs global futures that reject the exploitation of both humans and the natural world for endless wealth accumulation. This approach recognizes that environmental well-being is interconnected with social harmony. It involves collaborative initiatives that respect traditional ecological practices, incorporating them into conservation strategies. Decolonization also involves adopting a cosmological perspective of the Earth community as interdependent (Grim 2014), where humans are not seen as different from nature. For Indigenous people, it is possible to have a "world of many worlds" (Stenmark 2022), a pluriversal seascape and spiritscape, a maritime cosmos. Seeing place as a "co-inhabited cosmos" (Watson 2014, p. 93), as the Sama-Bajau communities do, has parallels to the Indigenous concepts of *buen vivir* or *sumak Kawsay* in Latin America (Kowii 2014) and Ubuntu in Africa (Brás 2021). Considering Indigenous conceptions of place and time matters because the social construction of these concepts can affect environmental peacebuilding processes and outcomes.

Second, related to the above, account for relational and alternative ways of knowing, being, and becoming. The Sama-Bajau look to the moon, stars, and weather to determine their fortunes or the best time for fishing (Solis 2022). They celebrate *pag-umboh* or the communal reverence towards their ancestors' spirits. From their cosmological lens, *Umboh Tuhan* is the most supreme being, residing everywhere, including the heaven and sea (Bottignolo 1995; Jubilado 2010; Sather 2001). The spirituality of Sama-Bajau plays a role in their understanding of resource management and fish abundance. They follow the tradition of *pamali* (taboo) and the consequences of violating it. For example, the disposal of fruit and vegetable peels overboard is seen as *pamali* as these land species do not belong in the sea. Different places have different prohibitions. Given that the sea has agency, is unpredictable, and that different marine species can appear in and out of place, the *pamali* local wisdom is a self-conscious behavior that ensures the sustainability of the seas. It leads to the conservation of marine resources, regulates fishing practices, protects coral reefs, and preserves marine animals seen as embodiments of *Umboh/Mbo'*, the sea deity (Andaya 2017; Lowe 2003; Maglana 2016). When adhering to the taboo, fishers are hopeful for higher catches in certain areas with favorable sea conditions. Failure to respect *pamali* leads to fear of perilous seas and reduced catch. Their belief in the sea deity and their perception of the sea as a sacred place shape their attitudes towards marine resources and their sense of responsibility in preserving them. The *pamali* system creates hope and fear, guiding individual actions and agency. Many places in the maritime landscape where *pamali* is upheld find instances of higher biodiversity (Basri et al. 2017; Simonin 2015). Environmental and spiritual factors help define community resilience, and understanding this leads to informed environmental peacebuilding policy and practice.

Amphibious thinking implies that inhabiting the land means inhabiting the water and vice-versa (Lowe 2003). Sea and land intertwine in Sama-Bajau marine worlds, similar to mangroves that straddle this interconnected space, with roots in saltwater while also reaching down into the soil of the coast. These Indigenous people view themselves as sharing "spiritual kinship and companionship" with the sea turtle because both are migratory and amphibious (Pauwelussen and Swanson 2022, p. 259). Adopting amphibious thinking—akin to seeing mangroves and sea turtles as living in a shared environment with littoral and terrestrial zones—is useful as it embraces flexibility and dispenses with the need to "fix" the Other through a "superior" Western development model (Pauwelussen and Verschoor 2017). Seeing global futures from this expanded lens aids in a culturally appropriate and responsive environmental peacebuilding theory and practice.

Fourth, similarly, decolonize the "victim" through Indigenous storytelling that considers their voices, histories, decision-making, and customary laws. An alternative narrative to their victimhood is to explore their agency, cultural identity, and resilience (Andal 2023). The dawn of the colonial market-based economy started the initial process of sedentarism. The arrival of more efficient fishing technology incentivized their move closer to the land, where their produce is in demand. Rather than seeing this as something negative, it can be viewed as flexibility and adaptation to shifting external socio-economic conditions. After all, having a fixed Sama-Bajau settlement does not mean they are permanently settled, as sedentary groups can resume nomadic lifestyles again. They should be redefined as "modern" subjects and their mobility should be considered outside developmentalist perspectives (Clifton 2014; Sopher 1965).

Fifth, decolonize by adopting sustainability and pluriversal peacebuilding (Karmakar and Chetty 2023). If one is to achieve a sustainable sea community, it is crucial to recognize and value the different dimensions of human-sea relationships beyond market considerations (material dimensions). Social and cultural values, a sense of belonging and place (relational dimensions), and personal fulfillment (subjective dimensions) can shape the well-being of the seas. Diverse perspectives and multiple stakeholder interests must be considered for a holistic and inclusive sea governance approach (Allison et al. 2020).

Sustainability benefits from embracing a care-oriented Indigenous perspective on the environment as a way of understanding and addressing ecological challenges (Watson 2014). This shift enables researchers to consider environmental peacebuilding as honoring the agency, spirituality, culture, and traditional knowledge of Indigenous people who are deeply embedded in the ecosystems they inhabit. Having a care-oriented Indigenous perspective delivers justice, dignity, well-being, and solidarity with the oppressed. It also allows for the inclusion of values and practices promoting environmental stewardship across generations, cultures, species, and territories. Indigenous lifeways and worldviews are not simply about using the environment as a resource; rather, it is about nurturing the Earth community (Berry 1988; Grim 2014) by engaging in reciprocal relationships, creative cosmological interactions, and finding ways towards sustainable futures on Earth.

In the present climate crisis, Indigenous people's experience and traditional ecological knowledge must be incorporated into global climate change policies. Climate is a cosmopolitical concern (Latour 2011)—addressing which benefits not only humans but also the cosmos, sustaining the physical and the metaphysical. There are Indigenous epistemologies on climate change, such as respecting nature's rhythms, that offer valuable lessons on sustainability, resilience, and the need for inclusive, culturally informed climate solutions (Watt-Cloutier 2018). Food insecurity brought by climate change is mitigated through reciprocal catch-sharing among the Sama-Bajau (Clifton 2014), which other societies can model on. Proactive steps are essential to preserve this wisdom and foster cross-cultural collaboration on climate solutions that respect, honor, and empower Indigenous peoples and heritage.

Sixth, decolonize from the orthodoxy of the "white male" bias underpinning the coloniality of gender (Lugones 2013). One can implode the hegemonic borders of Western heteropatriarchal epistemology by providing space for the localized knowledge of Indigenous feminist voices (Green 2020; Todd 2016). Indigenous women's epistemes should be platformed without fitting their experiences, histories, and voices to mainstream (Western) feminist frameworks (Sunseri 2007). The Sama-Bajau have distinct socio-economic and cultural dynamics: egalitarian structures and uxorilocal (matrilocal) practices characterize their households, forming matrilinear networks that facilitate information sharing, mutual support, and refuge during stressors. Sama-Bajau women play both leadership and support roles (Clifton 2014; Sather 1997; Stacey et al. 2017). They foster networks and preserve environmental knowledge across generations (La Ola et al. 2020). Thus, excluding them from decision-making restricts the communities' adaptive capacity to stressors; meaningfully including them aids in marine conservation success and addressing climate change impacts on marine food security (Ariando 2021; Stacey et al. 2012).

A decolonized environmental peacebuilding must account for the "pluriversality of context, history, experience, epistemology and ontology" (Suffla et al., 2020, p. 343). There are varied contexts and multiple alternatives to the Western development project—contemporaneous and overlapping worlds and realities—as well as ways of knowing and being. Dominant narratives must be re-evaluated in favor of a more just and diverse worldview, breaking ties with the colonial matrix of power inherent in Western modernity. In so doing, one can pave the path for what the Zapatistas describe as "a world in which many worlds can fit" (Maldonado-Villalpando et al., 2022).

## 6   Conclusion

The coloniality of power impedes Indigenous people from realizing their full potential by perpetuating systems that marginalize, exploit, and subjugate them. Colonial influences have characteristically defined environmental peacebuilding theory and practice, including the liberal peace agenda, positivism (the "universal" and objective value of science), and Western conservation models that exclude Indigenous participation. The case of the Sama-Bajau illustrates the challenges and opportunities that decolonizing environmental peacebuilding presents. The Sama-Bajau communities embody the complexities of balancing tradition and adaptation in the face of environmental changes, as well as addressing historical injustices and the different types of violence they face. By examining their experiences, this chapter provides valuable insights into how Indigenous communities negotiate their place in a world that forces them to the peripheries, exacerbated by a climate crisis—and the potential for decolonizing practices to empower and protect such communities.

Environmental peacebuilding is not expected to solve all issues; it is but a first step toward broader forms of cooperation. Nevertheless, it is relevant to bring to the fore the cultural and Indigenous approaches of the Sama-Bajau, the incompatibility between their traditional life-worlds and the Westphalian state system, and the evident power asymmetry because environmental issues often intertwine with these. Providing context is crucial here as every peacebuilding situation is unique, requiring an understanding of local dynamics, history, and culture to develop effective and sustainable peacebuilding strategies that address specific challenges in specific contexts. This is a starting point for discussing the issues.

**Acknowledgements** I recognize and honor the Indigenous people of maritime Southeast Asia whose deep connections to the sea, rich cultural traditions, resilience in the face of historic injustice, and profound wisdom have inspired this research. I am also grateful to Prof. Afshin Akhtar-Khavari (Queensland University of Technology), Prof. Sedfrey Candelaria (Ateneo de Manila University School of Law and Philippine Judicial Academy), Prof. Morgan Brigg (University of Queensland), Dr. Arthur Barraza (Griffith University), Shai Lagarde, and the anonymous peer reviewers for their thoughtful feedback and constructive criticisms. My gratitude likewise goes to Prof. Tobias Ide

(Murdoch University) and Giselle Miole (PhD student, Waseda University) as well as the conveners, audiences, practitioners, and researchers at the NERPS 2024 Conference in Japan for their insightful engagement with my chapter during my panel presentation.

# References

Abdi AA (2009) Oral societies and colonial experiences: Sub-Saharan Africa and the de-facto power of the written word. In: Education, decolonization and development. Brill, pp 39–56

Acciaioli G, Brunt H, Clifton J (2017) Foreigners everywhere, nationals nowhere: exclusion, irregularity, and invisibility of stateless Bajau Laut in Eastern Sabah, Malaysia. J Immigr Refug Stud 15(3):232–249

Acciaioli G (2001) 'Archipelagic culture' as an exclusionary government discourse in Indonesia. Asia Pac J Anthropol 2(1):1–23

Allerton C (2014) Statelessness and the lives of the children of migrants in Sabah, East Malaysia. Tilburg Law Rev 19(1–2):26–34

Allison EH, Kurien J, Ota Y (2020) The human relationship with our Ocean Planet. World Resources Institute

Amat A, Samad LA (2013) Bajau's Tekodon: a preliminary study. J Educ Soc Res 3(7):445

Andal AG (2023) Nomadic boat-dwelling children of Southeast Asia: discourses on the Sama-Bajau children and implications on decentring child migration studies. Child Soc 1–17

Andaya BW (2017) Seas, oceans and cosmologies in Southeast Asia. J Southeast Asian Stud 48(3):349–371

Andaya BW (2019) Recording the past of "peoples without history": Southeast Asia's sea nomads. Asian Rev 32(1):5–33

Aoyama W (2012) Social inequality among Sama-Bajau Migrants in Urban Settlements: a case from Davao City. Hakusan Rev Anthropol 15:7–44

Ariando MW (2021) Developing a model for the integration of Bajau traditional ecological knowledge in the management of locally managed marine area: a case study of Wakatobi Regency, Chulalongkorn University, Indonesia

Barrena J, Harambour A, Lamers M, Bush SR (2022) Contested mobilities in the maritory: implications of boundary formation in a nomadic space. Environ Plann c: Polit Space 40(1):221–240

Basri LO, Mudana IW, Habsah WOS, Marhadi A, Tarifu L, Burhan F, Janu L (2017) *Pamali*, Bajo's local wisdom in the conservation of marine resources. Asian Soc Sci 13(12):63–67

Berry T (1988) The dream of the earth, vol 2. Sierra Club Books

Borrows J (2016) Heroes, tricksters, monsters, and caretakers: Indigenous law and legal education. McGill Law J 61(4):795–846

Bottignolo B (1995) Celebrations with the sun: an overview of religious phenomena among the Badjaos. Ateneo de Manila University Press

Brás JGV (2021) For an epistemic decolonisation of education from the ubuntu philosophy. Pedag Cult Soc 1–16

Bellina B, Blench R, Galipaud J-C (2021) Sea nomadism from the past to the present. In Sea nomads of Southeast Asia: from the past to the present. NUS Press, pp 1–27

Bruch C, Jensen D, Emma M (2021) Defining environmental peacebuilding. In: Routledge handbook of environmental security. Routledge, pp 9–25

Brunt H (2013) Stateless Stakeholders: seen but not heard? The case of the Sama Dilaut in Sabah, Malaysia. University of Sussex, United Kingdom, Brighton

Butterly L, Richardson BJ (2016) Indigenous peoples and saltwater/freshwater governance. Indigen Law Bull 8(26):3–8

Carius A (2006) Environmental peacebuilding. Environmental cooperation as an instrument of crisis prevention and peacebuilding: conditions for success and constraints. Adelphi Report 3(7)

Castro-Gómez S (2019) The social sciences, epistemic violence, and the problem of the "invention of the other". In: Unbecoming modern. Routledge, pp 211–227

Chen C (2013) Mapping waters: thinking with watery places. In Chen C, MacLeod J, Neimanis A (eds) Thinking with water. McGill-Queen's University Press, p 274

Chen C, MacLeod J, Neimanis A (2013) Introduction: toward a hydrological turn? In Chen C, MacLeod J, Neimanis A (eds) Thinking with water. McGill-Queen's University Press, p 3

Chia CE (2016). Nomadic marginalities: the case of Bajau Laut's status within states and local economies in Semporna, Malaysia. MPA Diss., Central European University. Budapest, Hungary

Chiew H (2019) Bajau Laut: once sea nomads, now stateless. Retrieved 30 November 2023 from https://www.malaysiakini.com/news/476595

Clifton J (2014) Maritime ecocultures: Bajau communities of eastern Indonesia. In: Ecocultures. Routledge, pp 27–43

Clifton J, Majors C (2012) Culture, conservation, and conflict: perspectives on marine protection among the Bajau of Southeast Asia. Soc Nat Resour 25(7):716–725

Collins J, Thompson WK (2020) Aboriginal owned and jointly managed national parks: caring for cultural imperatives and conservation outcomes. In Rethinking wilderness and the wild. Routledge, pp 87–104

Collins YA, Macguire-Rajpaul V, Krauss JE, Asiyanbi A, Jiménez A, Bukhi Mabele M, Alexander-Owen M (2021) Plotting the coloniality of conservation. J Polit Ecol

Conca K, Dabelko GD (2002) Environmental peacemaking. Woodrow Wilson Center Press

Cooke FM, Acciaioli G (2021) Positioning Bajau identities as Bumiputera: challenges and potentials of leveraging environmental justice and Espousal of Islam in Sabah, Malaysia. In: Discourses, agency and identity in Malaysia: critical perspectives. Springer Singapore, vol 13, pp 153–181

Cusicanqui SR (2012) Ch'ixinakax utxiwa: a reflection on the practices and discourses of decolonization. South Atlantic Q 111(1):95–109

Dodds K, Kuus M, Sharp J (2013) The Ashgate research companion to critical geopolitics. Ashgate

Do Khac AL (2018) The political sea: conservation policies, state power, and symbolic violence: the case of the Bajau in the Wakatobi Marine National Park. Explor Grad Student J Southeast Asian Stud 14(Special Issue: Water in Southeast Asia):37–50

Dresse A, Fischhendler I, Nielsen JØ, Zikos D (2019) Environmental peacebuilding: towards a theoretical framework. Coop Confl 54(1):99–119

Elvey A (2020) Reimagining decolonising praxis for a just and ecologically sustainable peace in an Australian context. In: Towards a just and ecologically sustainable peace: navigating the great transition, pp 275–295

Famoso JMG (2021) Sama culture and Folk Narratives structures. Southeastern Philippines J Res Dev 26(2):83–96

Ferguson CE, Marie Green K, Switzer Swanson S (2022) Indigenous food sovereignty is constrained by "time imperialism." Geoforum 133:20–31

Galtung J (1969) Violence, peace, and peace research. J Peace Res 6(3):167–191

Garland E (2008) The elephant in the room: confronting the colonial character of wildlife conservation in Africa. Afr Stud Rev 51(3):51–74

Gaynor JL (2010) Flexible fishing: gender and the new spatial division of labor in Eastern Indonesia's rural littoral. Radic Hist Rev 2010(107):74–100

Gibbs LM (2010) "A beautiful soaking rain": environmental value and water beyond Eurocentrism. Environ Plann d: Soc Space 28(2):363–378

González Casanova P (1965) Internal colonialism and national development. Stud Comp Int Dev 1:27–37

Green J (2020) Making space for Indigenous feminism. Fernwood Publishing

Grim J (2014) Thomas Berry and indigenous thought: first nations and communion with the natural world. In: Eaton H (ed) The intellectual journey of Thomas Berry: imagining the earth community. Lexington Books, pp 123–147

Hanasz PM (2016) An examination of the South Asia water initiative and associated donor-led processes in the transboundary water governance of the Ganges-Brahmaputra problemshed

Hardt JN, Scheffran J (2019) Environmental peacebuilding and climate change: peace and conflict studies at the edge of transformation. Toda Peace Institute Policy Brief, vol 68, pp 1–20

Hoogervorst TG (2012) Ethnicity and aquatic lifestyles: exploring Southeast Asia's past and present seascapes. Water Hist 4:245–265

Hunt S (2014) Ontologies of indigeneity: the politics of embodying a concept. Cult Geogr 21(1):27–32

Harris I, Mische P (2008) Environmental peacemaking, peacekeeping, and peacebuilding. In Bajaj M (ed) Encyclopedia of peace education. Columbia University

Ibrahim WZ (2021) The new economic policy and contesting Bumiputera identity. Institute for Democracy and Economic Affairs (IDEAS) Malaysia

Ide T (2019) Environmental peacemaking and environmental peacebuilding in International politics. University of Hamburg, Research Group Climate Change and Security

Ide T, Bruch C, Carius A, Conca K, Dabelko GD, Matthew R, Weinthal E (2021a) The past and future (s) of environmental peacebuilding. Int Aff 97(1):1–16

Ide T, Palmer LR, Barnett J (2021b) Environmental peacebuilding from below: customary approaches in Timor-Leste. Int Aff 97(1):103–117

Illouz C, Nuraini C (2021) The Bajau Diaspora: origin and transformation. In Bérénice Bellina RB, Galipaud J-C (eds) Sea nomads of Southeast Asia: from the past to the present. NUS Press

Jegen LF (2023) 'Migratising' mobility: coloniality of knowledge and externally funded migration capacity building projects in Niger. Geoforum 146:103862

Jubilado RC (2010) On cultural fluidity: the Sama-Bajau of the Sulu-Sulawesi Seas. KUNAPIPI: J Postcolonial Writ Cult 32(1&2):89–101

Kahn-John M, Koithan M (2015) Living in health, harmony, and beauty: the Diné (Navajo) Hózhó wellness philosophy. Glob Adv Health Med 4(3):24–30

Karmakar G, Chetty R (2023) Tackling environmental and epistemic injustice: decolonial approaches for Pluriversal peacebuilding in South Africa. Peace Rev 1–15

Koopman S, Dalby S, Megoran N, Sharp J, Kearns G, Squire R, Jeffrey A, Squire V, Toal G (2021) Critical Geopolitics/critical geopolitics 25 years on. Polit Geogr

Kowii A (2014) El Sumak Kawsay. In: Mosquera GH (ed) Antología del Pensamiento Crítico Ecuatoriano Contemporáneo. Pydlos, CIM, FIUCUHU, pp 437–444

Kyrou CN (2007) Peace ecology: an emerging paradigm in peace studies. Int J Peace Stud 73–92

La Ola T, Wianti N, Tadjuddah M, Buana T, Abdullah S, Wunawarsih I (2020) Social network analysis: key actors of cooperation between Sama Bajo and land-dwellers in Wakatobi marine national park. In: IOP conference series: earth and environmental science

Latour B (2011) Politics of nature: east and west perspectives. Ethics Glob Polit 4(1):71–80

Lowe C (2003) The magic of place: Sama at sea and on land in Sulawesi, Indonesia. Bijdragen tot de Taal-, land-en Volkenkunde 159(1):109–133

Lugones M (2013) The coloniality of gender. In: Globalization and the decolonial option. Routledge, pp 369–390

Mac Ginty R (2008) Indigenous peace-making versus the liberal peace. Coop Confl 43(2):139–163

Macas L (2010) Sumak Kawsay: La vida en plenitud. América Latina En Movimiento 452:14–16

Maglana MC (2016) Understanding identity and diaspora: the case of the Sama-Bajau of Maritime Southeast Asia. Jurnal Sejarah Citra Lekha 1(2):71–80

Maldonado-Villalpando E, Paneque-Gálvez J, Demaria F, Napoletano BM (2022) Grassroots innovation for the pluriverse: Evidence from Zapatismo and autonomous Zapatista education. Sustain Sci 17(4):1301–1316

Marshall S, Mahali SNH, Januin J (2019) Contestations over Malaysian Citizenship and the Preliminary Case for Recognising the Stateless Bajau Laut Community. Borneo Res J 66–78

McNiven I (2004) Saltwater people: spiritscapes, maritime rituals and the archaeology of Australian indigenous seascapes. World Archaeol 35(3):329–349

Medina MCT (2016) Aftermath of the 2013 Zamboanga Siege: peacebuilding strategies from civil society. Philippine Sociol Rev 9–39

Mezzadra S, Neilson B (2017) On the multiple frontiers of extraction: excavating contemporary capitalism. Cult Stud 31(2–3):185–204

Mignolo WD (2011) The darker side of western modernity: global futures, decolonial options. Duke University Press

Moreton-Robinson A (2009) Imagining the good indigenous citizen: race war and the pathology of patriarchal white sovereignty. Cult Stud Rev 15(2):61–79

Moreton-Robinson A (2021) The white possessive: identity matters in becoming Native, Black and Aboriginal. Borderlands Journal 20(2):4–29

Neezzhoni DZ (2010) Diné education from a Hózhó perspective [MA Diss., Arizona State University]. Phoenix

Nimmo HA (1968) Reflections on Bajau history. Philipp Stud 16(1):32–59

Nor MAM (2017) Perspectives on Bajau/Sama' Diaspora. In: Sabah Museum monograph. Department of Sabah Museum, vol 13

North R (2017) Traditional fishing rights in international law: 'The South China sea arbitration.' Univ Tasmania Law Rev 36(1):101–108

Nuraini C (2016) The Intangible Legacy of the Indonesian Bajo. Wacana, J Human Indonesia 17(1):1–18

Paris R (2002) International peacebuilding and the 'mission civilisatrice.' Rev Int Stud 28(4):637–656

Pauwelussen A (2016) Community as network: exploring a relational approach to social resilience in coastal Indonesia. Marit Stud 15(1):2

Pauwelussen A, Swanson SS (2022) Good human–turtle relationships in Indonesia: exploring intersecting legalities in sea turtle conservation. In: Laws of the sea. Routledge, pp 258–281

Pauwelussen AP, Verschoor G (2017) Amphibious encounters: coral and people in conservation outreach in Indonesia. Engag Sci Technol Soc 3:292–314

Quijano A (2000) Coloniality of power and Eurocentrism in Latin America. Int Sociol 15(2):215–232

Quijano A (2007) Coloniality and Modernity/rationality. Cult Stud 21(2–3):168–178

Rahmawati R, Kahirun K (2020) Impact of climate change and community adaptation of Bajo Tribe. J La Lifesci 1(2):1–6

Richmond OP, Mac Ginty R (2015) Where now for the critique of the liberal peace? Cooper Conflict 50(2):171–189

Roxas-Lim A (2017) Marine-oriented Sama-Bajao people and their search for human rights. Public Policy 18:49–66

Sather C (1997) The Bajau Laut: adaptation, history, and fate in a maritime fishing society of South-East Sabah. Oxford University Press

Sather C (2001) Bajau Laut boat-building in Semporna. In: Techniques & culture. Revue semestrielle d'anthropologie des techniques, vols 35–36, pp 177–198

Sather C (2003) Keeping the Peace in an Island World: the Sama Dilaut of Southeast Asia. In Kemp G, Fry DP (eds) Keeping the peace: conflict resolution and peaceful societies around the world. Taylor & Francis Group, p 101

Schirch L (2022) Decolonising peacebuilding. In Berghof handbook for conflict transformation. Berghof Foundation, in collaboration with Toda Peace Institute, Berlin

Searle T, Muller S (2019) "Whiteness" and natural resource management: let's talk about race baby, let's talk about sovereignty! Geogr Res 57(4):411–424

Simon SE, Mona A (2023) Between legal indigeneity and Indigenous sovereignty in Taiwan: insights from critical race theory. Soc Inclus 11(2):187–197

Simonin PW (2015) From sea to spirit: resilience conceptions in coastal communities of Kaledupa, Indonesia. Resilience 3(3):199–206

Solis J (2022) Ang konsepto sa kalluman ng Badjao: Pagsilang hanggang pagkamatay. Int J Res 11(14):59–77

Somiah V (2022) The sea is indigenous 'land' too: negotiating presence and rights of indigenous maritime communities in Sabah, Malaysia. Sojourn: J Soc Iss Southeast Asia 37(1):85–112

Somiah V, Santo Domingo JR (2021) Sabah's unrelenting exclusionary and inclusionary politics. In: S. Rajaratnam School of International Studies (RSIS) Working Papers. Nanyang Technological University, Singapore

Sopher DE (1965) The sea nomads: a study based on the literature of the maritime boat people of Southeast Asia. Syracuse University, Memoirs of the National Museum, vol 5

South China Sea Arbitration (2016) In Philippines v China, Award. Permanent Court of Arbitration (PCA) No. 2013-19, ICGJ 495 (PCA 2016), 12 July 2016

Srigley K, Sutherland L (2018) Decolonizing, indigenizing, and learning Biskaaybiiyang in the field: our oral history journey. Oral Hist Rev 45(1):7–28

Stacey N (2007) Boats to burn: Bajo fishing activity in the Australian fishing zone. ANU Press

Stacey, NE, Karam J, Meekan MG, Pickering S, Ninef J (2012) Prospects for whale shark conservation in Eastern Indonesia through bajo traditional ecological knowledge and community-based monitoring. Conserv Soc 10(1):63–75

Stacey N, Acciaioli G, Clifton J, Steenbergen DJ (2017) Impacts of marine protected areas on livelihoods and food security of the Bajau as an indigenous migratory people in maritime Southeast Asia. In: Westlund L, Charles A, Garcia SM, Sanders J (eds) Marine protected areas: interactions with fishery livelihoods and food security. UN Food and Agriculture Organisation, pp 113–126

Stavrevska EB, Carvajal SZ, Luci N (2022) 'Peace' across spaces: discussing feminist (and) decolonial visions of peace. In: Feminist conversations on peace. Bristol University Press, pp 34–46

Stenmark LL (2022) Telling Stories in the Pluriverse: decolonial options for creative pluralism. In: Issues in science and theology: creative pluralism? Images and models in science and religion. Springer, pp 45–57

Suffla S, Malherbe N, Seedat M (2020) Recovering the everyday within and for decolonial peacebuilding through politico-affective space. In: Acar YG, Moss SM, Uluğ ÖM (eds) Researching peace, conflict, and power in the field: methodological challenges and opportunities. Springer International Publishing, pp 343–364

Sunseri L (2007) Indigenous voice matters: claiming our space through decolonising research. Junct: J Themat Dialogue 9

Tahara T (2021) The business network of Bajau Tribe Sea fisheries on the Indonesia Malaysia Border. Acad Entrepren J 27(1)

Teschke B, Lacher H (2007) The changing 'logics' of capitalist competition. Camb Rev Int Aff 20(4):565–580

The British North Borneo Herald (1886) citing a Company Officer's District Report (1 October 1886), p 221

Todd Z (2016) An indigenous feminist's take on the ontological turn: 'ontology' is just another word for colonialism. J Hist Sociol 29(1):4–22

Tuathail GÓ, Toal G (1996) Critical geopolitics: the politics of writing global space (Vol. 6). U of Minnesota Press

UNHCHR Philippines (2019) Ending statelessness for the next generation. Retrieved 30 September 2024 from https://www.unhcr.org/ph/16771-october-e-newsletter-statelessness.html

UNICEF Philippines (2024) Over 1,300 at risk of statelessness receive birth certificates in Maguindanao and BaSulTa. Retrieved 30 September 2024 from https://www.unicef.org/philip pines/press-releases/over-1300-risk-statelessness-receive-birth-certificates-maguindanao-and-basulta

Watson MC (2014) Derrida, Stengers, Latour, and subalternist cosmopolitics. Theor Cult Soc 31(1):75–98

Watt-Cloutier S (2018) The right to be cold: one woman's fight to protect the Arctic and save the planet from climate change. U of Minnesota Press

Whyte KP (2019) Way beyond the lifeboat: an indigenous allegory of climate justice. Clim Fut Reimag Glob Clim Justice 11–20

Ziegler DB (2022) The shameful statelessness of South-East Asia's sea nomads. Retrieved 30 November 2023 from https://www.economist.com/asia/2022/10/13/the-shameful-statelessness-of-south-east-asias-sea-nomads

# Leveraging Technology and Collaboration for Peacebuilding: A Case Study of Aretes Style Empowering Bangsamoro Women in Lanao Del Sur, Philippines

Yasmin A. Tagorda and Yasmin M. Mangotara

**Abstract** This chapter explores the potential of digital transformations and collaborative partnerships in promoting transformation within Aretes Style, a youth-led social enterprise. It details the interconnected challenges of peacebuilding, sustainability, and the empowerment of marginalized post-conflict communities. Using a mixed-method research design, qualitative and quantitative research methodologies were combined to evaluate the significant influence of digital transformations on income creation, community resilience, and peace advocacy in the vulnerable villages of Lanao del Sur. Furthermore, this chapter examines the co-production tactics employed by Aretes Style, focusing on their capacity to leverage the expertise and artistic capabilities of Sagonsongan internally displaced artisans and Tugaya weavers. The results highlight the notable impact of Aretes Style's inclusive methodology, which amplifies economic prospects for internally displaced persons (IDPs) and empowers them to engage in community development actively. The exploration emphasizes the significance of forming partnerships and fostering collaboration among diverse stakeholders, such as local communities, development organizations, and social entrepreneurs, to effectively attain favorable results in peacebuilding and socio-economic development. As woven by a tale of transformation and resilience, Aretes Style appears a symbol of optimism, showcasing the potential for social enterprises to effectively address the economic requirements of marginalized communities, facilitate social change, and contribute to enduring peace and sustainability through technology, cultivation of partnerships, and collaboration.

Y. A. Tagorda (✉)
De La Salle University, Manila, Philippines
e-mail: yasmin_tagorda@dlsu.edu.ph

Y. A. Tagorda · Y. M. Mangotara
Mindanao State University, Marawi, Philippines

Y. M. Mangotara
University of the Philippines Diliman, Quezon City, Philippines

© The Author(s), under exclusive license to Springer Nature Singapore Pte Ltd. 2024
A. Sharifi et al. (eds.), *Navigating Peace and Sustainability in an Increasingly Complex World*, World Sustainability Series, https://doi.org/10.1007/978-981-97-8772-2_4

**Keywords** Digital transformations · Post-conflict peacebuilding · Collaboration · Social enterprise · Marawi

# 1 Introduction

The backdrop of the Marawi Siege in 2017, a devastating conflict that significantly disrupted the socioeconomic fabric of Marawi City, sets the stage for this chapter. The siege, rooted in the historical grievances of the Moro people and fueled by extremist ideologies, notably by the Islamic State of Iraq and Syria (ISIS), transformed Marawi into a war zone and exacerbated poverty and instability in the region. It displaced over 350,000 individuals and resulted in an estimated PHP 11 billion (approximately USD 220 million) in damages, underscoring the critical need for comprehensive recovery and reconciliation efforts (Banlaoi 2020; Ordinario 2017; World Bank 2018). Economic growth and development are recognized as pivotal for conflict resolution and the establishment of peaceful societies, where alleviating poverty and inequality reduces the likelihood of conflict escalation (Collier 2007; Sachs 2005; Sen 1999).

However, to bridge the gap between the immediate crisis and long-term sustainability, Sharifi et al. (2023) advocate for an integrated approach that transcends traditional boundaries. Their insights suggest that in post-conflict contexts, more is needed to address economic and social needs in isolation. Instead, a multidisciplinary strategy of collaborative efforts and innovative policymaking can pave the way for enduring peace and prosperity.

This is the case of Marawi City, which has witnessed the implementation of numerous initiatives aimed at fostering recovery and development. For instance, the government and non-governmental organizations have been reconstructing infrastructure and providing psychological support to trauma victims. Educational initiatives aimed at bolstering community resilience have received financial support from international organizations and local cooperatives, both of which serve to promote economic activity (Cornelio and Calamba 2022).

The youth-led Aretes Style, a social enterprise, contributes to efforts toward sustainable peacebuilding by addressing the economic and social needs of Marawi's displaced population through co-production and collaboration, coupled with technological engagement. Echoing the sentiment "we are artisans, not terrorists," Aretes Style emerged as a transformative force to shift the narrative from violence to cultural and economic resilience by leveraging the artisanal skills of displaced and marginalized women in the community.

The literature review anchors the study in a web of interrelated themes—economic empowerment, social reconciliation, and the utilization of digital resources—integral to sustainable development and conflict resolution. Existing studies have highlighted the significant contribution of social enterprises in war-torn regions (Aldairany et al. 2018; Maracine 2019; Sottini and Ciambotti 2021), alongside the role of digital platforms in enhancing value creation and fostering networks (Fu et al. 2021; Battisti

2019; Devarapalli and Figueira 2015). The importance of networks and collaboration is underscored in the literature, revealing the role of relationships and social capital in the prosperity of social enterprises (Firchow 2018; Hidalgo et al. 2021; Larrauri and Kahl 2013), while also considering the potential challenges of digital technology integration (Hirblinger et al. 2023; Larrauri and Kahl 2013).

The literature investigation has revealed a significant gap in the actual application of knowledge within youth-led social enterprises that advocate for peacebuilding initiatives. Youth-led social enterprises are playing an increasingly important role in post-conflict peacebuilding and economic development, which have received little attention. We aim to fill this gap in this chapter by investigating Aretes Style's initiatives to promote peace and development through co-production and collaborative approaches, providing vital insights into their potential for long-term peacebuilding and economic resilience.

In this chapter, we explore this overarching question: How has Aretes Style leveraged technology and collaboration to promote peacebuilding, empowerment, and sustainability in post-conflict Marawi City? Specifically, it seeks to know: What roles and functions does Aretes Style play in post-conflict peacebuilding and community development? How does the enterprise manage and optimize its relationships and dynamics with networks and collaborators to advance its goals? Furthermore, the study investigates how Aretes Style utilized the digital landscape, including specific social media platforms, to fulfill its peacebuilding and economic empowerment objectives, and what challenges and obstacles it faces in these pursuits.

This case study provides a different perspective on the merging of technology and collaboration by considering how a social enterprise uses social media platforms to enhance their commercial success and advocate for peacebuilding. It deepens our knowledge about how social enterprises in post-conflict settings can exploit digital platforms to facilitate peacebuilding and economic empowerment, employing both qualitative data and social media analytics. Specifically, the study presents a new methodological design, which combines social media platform metrics such as likes, shares, and comments, vis-à-vis follower count, intended to project reach and engagement. However, the exploration only involved metrics available to the public, hence any specific data only viewable through user metrics such as monthly reports were not utilized.

By addressing these key areas, the research contributes to peacebuilding theory and practice, especially youth-led social enterprises during conflict and peacetime; the role of social media in peacebuilding and social entrepreneurship; as well as situations specific to the Philippines with a focus on the Marawi case study. The results provide insight into larger debates around empowerment, social entrepreneurship, and youth engagement that may inform theoretical implications. As such, other marginalized or post-conflict contexts can benefit from the successful integration of digital technology, collaborative partnerships, and cultural heritage preservation in generating economic empowerment and social cohesion. This all-inclusive nature emphasizes the wider importance of this work within discourses on empowerment and sustainable development, with practical takeaways for policymakers, development actors, or even social entrepreneurs aspiring to achieve long-term peaceful coexistence via creative and inclusive approaches.

## 2   Literature Review

### 2.1   The Interplay of Poverty and Conflict: How Social Enterprises Promote Economic Revival and Peace in Post-conflict Societies

The intricate link between poverty and conflict is crucial, as extensive research shows a two-way relationship where poverty not only results from conflict but also contributes to its occurrence. This concept, commonly referred to as the "conflict poverty trap," suggests that poverty can serve as both a symptom and a factor in conflict, disrupting economic activities and displacing communities, thus perpetuating poverty (Justino 2010). Poverty exacerbates conflict by making nations more susceptible to civil war, weakening governance and economic performance, and increasing the likelihood of conflict recurrence. On a national level, poverty weakens the ability to provide public goods and undermines institutions, making countries less resilient to conflict. Political oppression in semi-democratic regimes further highlights the complexity of poverty by intensifying tendencies toward violence (Bandyopadhyay and Younas 2011; Justino 2010). Effective governance and wise resource management can help reduce these risks, while political marginalization, inequality, and grievances based on identity add layers of complexity to the cycle (Humphreys and Weinstein 2008). This underscores the importance of reducing poverty and promoting economic development as essential pillars for preventing and resolving conflicts.

The World Bank's initiatives focus on strengthening resilient communities and fostering social unity through inclusive service delivery and citizen empowerment, which are particularly effective in fragile and conflict-affected areas by building trust and cohesion. Incorporating marginalized groups into economic programs is vital, as emphasized by the ICMA, which highlights the importance of developing strategies tailored to the specific needs and priorities of diverse communities (World Bank 2018; ICMA 2021).

Addressing the link between poverty and conflict calls for a holistic strategy that covers economic progress, political stability, and societal inclusiveness. Social enterprises stand out in business as entities with a purpose beyond profit. They are the harbingers of social innovation, leveraging entrepreneurial activities to advocate for societal betterment. Weerawardena and Sullivan Mort (2006) describe social entrepreneurs as change agents identifying and exploiting opportunities to initiate social transformation. In conflict-ridden areas, social enterprises act as vehicles for economic regeneration and conduits for peace, instilling a sense of normalcy and optimism in devastated communities (Sottini and Ciambotti 2021). This is substantiated by Maracine's (2019) study of Northern Uganda, where social enterprises have been instrumental in empowering local communities, especially women, fostering a grassroots movement towards reconciliation and trust-building.

Social enterprises are hybrid organizations that seek to harmonize commercial strategies with social missions. They are characterized by their steadfast commitment to addressing social issues while ensuring economic sustainability (Battilana et al. 2014). These organizations strike a balance between generating income and achieving impactful social change. The literature delineates social enterprises as entities that promote innovation and cultural cohesion, which are crucial in empowering marginalized groups and fostering community collaboration (Hidalgo et al. 2021; Larrauri and Kahl 2013).

## 2.2 Empowering Social Enterprises: The Interplay of Digital Technology, Social Capital and Institutions

Social enterprises are vital in post-conflict recovery and peacebuilding, integrating economic and social objectives to empower communities and foster reconciliation. Their use of technology and emphasis on cultural and social cohesion solidify their role as critical agents of change, leveraging digital tools to enhance operations and impact (Fu et al. 2021; Battisti 2019). However, applying these tools requires a focus on human-centered, context-specific solutions to ensure inclusivity and local relevance in development and peacebuilding efforts (Hirblinger et al. 2023).

Digital technologies are indispensable in sustaining social enterprises, providing innovative platforms for broader engagement and service delivery. These technologies, particularly social media, facilitate connections with diverse audiences, allowing enterprises to share their stories and advocate for causes effectively. Platforms like Facebook, Twitter, and Instagram have become crucial in storytelling, mobilizing collective action, building community support, and enhancing customer relationships and operational reach (Jones et al. 2015; Fu et al. 2021). Integrating digital tools allows for efficient scaling of sustainable solutions, though challenges such as the digital divide and value appropriation issues must be navigated carefully (Fish and Srinivasan 2011).

In peacebuilding, digital media offers a robust medium for broadcasting messages and galvanizing support. These platforms provide a global stage for advocacy and marketing, transcending traditional communication boundaries and fostering dialogue and support networks. The strategic use of social media can bolster value creation and drive stakeholder engagement, which is critical for the sustainability of peace-centric initiatives (Battisti 2019). However, integrating digital technology requires careful consideration to avoid perpetuating biases or exclusion, emphasizing context-specific strategies (Hirblinger et al. 2023; Larrauri and Kahl 2013).

Social media offers both opportunities and challenges in the field of peacebuilding. It can be a tool for advocating, mobilizing, and raising awareness about peace-related issues, connecting peace advocates worldwide. Conversely, it can also be misused to spread hate speech and provoke violence, hindering peacebuilding efforts. Some individuals may misuse media to spread malicious messages, target opponents with harassment, and plan disruptive actions that hamper the peace process (Majcin 2019; Fahmy and Hussain 2021). Additionally, factors like literacy and restricted access for certain groups can create an uneven playing field, amplifying some voices while marginalizing others, posing a threat to inclusive and fair peacebuilding processes. These obstacles underscore the importance of implementing rounded strategies when utilizing media for peacebuilding purposes (Majcin 2019; Fahmy and Hussain 2021). This involves monitoring hate speech, promoting literacy initiatives, and establishing policies that ensure these platforms are accessible and safe for all individuals, regardless of their background or status. By taking these measures, we can cultivate an online environment conducive to fostering peace and reconciliation initiatives (Conflict Management Consulting 2016).

Social capital, which encompasses resources, trust, and networks essential for facilitating mutual coordination and cooperation, is a fundamental factor in driving the success of social entrepreneurship (Hidalgo et al. 2021). The fostering of collective action and the acquisition of support are facilitated by its reliance on interpersonal ties and commonly held standards. Social entrepreneurs effectively leverage this capital by adeptly establishing interpersonal relationships, networks, and trust, which are crucial for garnering support from the community and mobilizing resources.

The influence of social capital extends beyond networking, encompassing the credibility and long-term viability facilitated by organizations such as cooperatives and non-governmental organizations (NGOs), which are fundamental components of the social capital framework. These entities validate and enhance social enterprises' resilience, enabling them to achieve their objectives efficiently.

Nevertheless, the domain of social entrepreneurship presents distinct problems and opportunities. Reconciling financial objectives with the social impact mission poses a multifaceted situation for social enterprises. The attainment of success within this particular domain is subject to various influential elements, such as empathy, self-efficacy, social networks, and the availability of resources. Moreover, the landscape of social entrepreneurship is significantly influenced by external factors such as government regulations, socioeconomic situations, and cultural norms. A combination of factors influences the creation and evolution of social entrepreneurship in different contexts. These factors underscore the interdependent connection between social capital and social entrepreneurship, with each factor mutually reinforcing the other (Hidalgo et al. 2021).

Institutional frameworks, including legal, socio-political, and cultural norms, significantly influence the efficacy of social enterprises. Collaborative partnerships between social enterprises, NGOs, governments, and private sectors, fostered by supportive institutional environments, are vital for spurring innovation and amplifying impact, thereby leading to sustainable development and peacebuilding (Hidalgo et al. 2021; Larrauri and Kahl 2013; Firchow 2018). However, challenges such as governmental interference and corruption can pose significant barriers to their development and sustainability (Maracine 2019).

The synergy between digital technology and robust institutional frameworks is vital for the success of social enterprises in peacebuilding and community development. While digital platforms offer innovative engagement methods and a global reach, supportive institutional environments provide the foundation for these enterprises to thrive and impact meaningfully.

## 2.3   Women in Peacebuilding: Empowering Women Through Social Enterprises

The link between women and peace is fundamentally based on the idea that women play a significant role in promoting peace and resolving conflicts. It is often argued that including women in peace processes is beneficial due to their unique experiences and perspectives, particularly in conflict-affected areas. Feminist peace theory suggests that empowering women and involving them actively can lead to more comprehensive and sustainable peacebuilding outcomes. This is partly because women are commonly viewed as promoters of social unity and may approach conflict resolution by emphasizing reconciliation and community development (Väyrynen 2010).

Research findings indicate that women play a unique and transformational role in peace processes. Their efforts are usually geared toward engaging citizens at the community level with grassroots initiatives designed to foster peaceful coexistence, trust among communities, and other social aspects of peace (Anderlini 2007). The involvement of women in peace processes provides diverse points of view about rights, justice, and the socio-economic interests of affected communities (Paffenholz 2016).

Anderlini (2007) and Porter (2008) affirm that women make a significant contribution to the peace process, and excluding women as essential stakeholders in peace negotiations reduces the likelihood of achieving long-term and inclusive peace. According to Porter (2008), women have been able to play the role of officials who can ease clan differences and act as informal negotiators; they have the potential to influence peace initiatives even when they are not official participants.

Shepherd (2016) posits that women's participation in peacebuilding alters its dynamics, challenging the ability and authority to define peacemaking work and power distribution within society. According to Porter (2008), women are more likely to deal with the root causes of conflict, such as gender equality, reconciliation, and nonviolence. They are critical in creating long-term recovered communities in post-conflict circumstances (Adjei 2019).

Women in Bangsamoro have helped to establish peace through lobbying, economic empowerment, health education, and communal activities. In particular, women-led organizations in the region, such as Noorus Salaam and Ummah Fi Salam-Women, have each dedicated their purpose towards building peace through particular objectives of peace education and livelihood support. For instance, Noorus Salam fosters peace through nonviolence, mutual tolerance, and respect, raising community awareness and education on peace and reconciliation. These activities foster communal harmony. On the other hand, Ummah Fi Salam-Women provides capacity training and microfinance to help women achieve economic empowerment. This support is necessary for both economic progress and cultural preservation (Tagorda and Magno 2023).

Engagement in entrepreneurship can help generate peace by empowering women to make constructive changes. Women can help improve economic stability, reduce poverty, and address inequalities by starting their own enterprises (Paffenholz 2016). Societies with stronger economies and more equitable systems are less likely to experience conflict (Tagorda and Magno 2023). Furthermore, social entrepreneurship allows women to take control of their lives while helping others, resulting in increased participation in decision-making processes and peacebuilding efforts. Empowering women through entrepreneurship increases their impact in their households, communities, and society as a whole. This empowerment enables them to actively advocate for peace by challenging conventions that produce conflict and violence, thereby creating environments that encourage the establishment of channels that contribute to sustainable peace.

## 3  Framework of Analysis

The success of a youth-led initiative working with women to empower them and ultimately contribute to fostering peace may vary depending on different institutional settings. By employing an institutional analysis, we can explore the factors influencing the involvement of Aretes Style in social entrepreneurship and peace efforts. This chapter highlights the role played by institutions in shaping their activities within the sociocultural, political, and economic environments of Lanao del Sur and the Bangsamoro Autonomous Region in Muslim Mindanao (BARMM).

Advocates of Institutionalism argue for the central role of institutions (Peters 2005; Schmidt 2006) by serving as mediators in influencing actors' behaviors and subsequent political outcomes (Hay 2002). They suggest that focusing on institutions rather than individuals is a valuable analytical approach. Moreover, there are varying perspectives among Institutionalists regarding how institutions should be defined and why they are important, and one comprehensive perspective outlined by North (1990, p. 1) describes institutions as "the created constraints that guide human interaction" or "the established rules, within a society." This definition encompasses key aspects: first, institutions are designed by humans to regulate conduct and shape incentives; second, they encompass various de facto and de jure aspects spanning economic, political, and social domains.

Hence, reiterating North's (1990) emphasis on how institutions shape behavior through a set of rules, norms, and expectations that impact behaviors and relationships, institutions as determinants of societal outcomes manifests in the context of institutions in the Bangsamoro region influencing individuals' decisions and choices for actors within its greater influence (e.g., Aretes Style, Noorus Salaam and Ummah Fi-Salam-Women).

To further emphasize how institutions in the Bangsamoro Region play a crucial role in shaping societal dynamics for key actors in the study, an exploration of how and where the political, cultural, and economic structures influence the successes and challenges faced by women and Aretes Style in their endeavors related to entrepreneurship and peacebuilding are drawn henceforth.

First, the political institutional context in the region involves examining the political systems, power dynamics, governance structures, and policies that define Aretes Style's role in social entrepreneurship and peacebuilding. It explores how opportunities for engagement in peacebuilding processes are influenced by legal frameworks, political beliefs, and interactions with governmental bodies (North 1990; Peters 2005; Schmidt 2006).

Second, the cultural landscape is another institutional element affecting collaborations between Aretes Style and female weavers in social enterprise projects. Understanding the values, beliefs, traditions, and identities of the communities they engage with is essential for the success of their initiatives. This aspect involves examining how the actors adjust their strategies for enterprise and peacebuilding, including messaging and community engagement methods, to align with cultural values and sensitivities.

Finally, the economic environment significantly influences their efforts to promote enterprise and peace in the region. This aspect examines the status of the community, disparities, distribution of resources, and opportunities for livelihood. It explores the impact of these challenges on Aretes Style's capability in fundraising efforts and sustainability in carrying out peace advocacy initiatives.

# 4   Methods

This chapter adopts a mixed-method research design to understand how Aretes Style navigates its peacebuilding objectives and efforts, such as income generation, empowering women, and fostering social cohesion, through the socio-political web of Meranaw communities and the broader context of the Philippines. The mixed-methods approach was selected for its ability to combine the richness of qualitative data with the objectivity and generalizability of quantitative descriptions, thereby providing a more comprehensive understanding of Aretes Style's initiatives.

The qualitative phase aims to capture the real-life experiences, challenges, and strategies employed by Aretes Style to reach its goals. This includes detailed semi-structured interviews, which probed individual narratives and perspectives to understand how participants navigate formal and informal structures, adhere to cultural norms, and utilize legal-political frameworks in pursuit of their advocacies. The use of focus group discussions (FGDs) brought together Tugaya weavers and Sagonsogan artisans for discussions that allowed for shared storytelling. This method uncovers challenges and strategies that may not surface in one-on-one interviews. The FGDs also explore how cultural norms and economic constraints influence their involvement in the enterprise and peacebuilding efforts. Key informant interviews (KIIs) involve actors such as government officials, representatives from non-governmental organizations (NGOs), donor agencies, and local leaders from the community and religious sectors. This approach gathers a range of views on how the institutional environment shapes the operations and community interactions of Aretes Style. We focus our questions on how policies and informal social norms influence Aretes Style's activities.

We analyzed our qualitative data using the software QDA Miner, supplemented by manual coding techniques, to arrive at themes for thematic analysis. QDA Miner facilitated the systematic organization and coding of large volumes of text, enhancing the efficiency and accuracy of identifying key patterns. Manual coding was employed to ensure a deeper understanding and contextual interpretation of the data. This combined approach enabled the extraction of relevant and substantial themes, which provided comprehensive insights into the reach of social media for advocacy purposes, highlighting the specific challenges and successes experienced by Aretes Style.

The quantitative phase of the study focuses on evaluating the effectiveness and extent of Aretes Style's use of media for marketing and advocacy purposes. This includes digital platform analytics by examining and interpreting publicly available data from platforms like Facebook, Instagram, and The Spark Project to measure audience reach and engagement levels related to the dissemination of messages about peacebuilding and community empowerment. "Effectiveness" in this sense was explored through "reach" and "engagement." The variety of platforms used (Facebook, Instagram, and The Spark Project) and their corresponding metrics add to reach. In a more specific sense, *likes* reflected on Aretes Style's social media platforms represent Facebook and Instagram reach, while *funding raised* represents the reach of The Spark Project. In particular, Jal Mustari emphasized that the purpose of Facebook and Instagram depends on certain audiences. While the enterprise's Instagram account was registered with the intention of attracting influencers and small businesses looking to promote, Facebook was for the general public (J. Mustari, personal communication, August 6, 2023). Moreover, the Spark Project, an online crowdfunding channel, was created for creative changemakers seeking crowdfunding and donors seeking network expansion and contribution to sustainable enterprises. Registration was also secured for this platform. Vis-à-vis the number of likes and views, follower count was also considered to measure engagement since engagement is detailed as the actual interaction of followers with posts, which in turn suggests a layer of contribution to advocacy as it engages followers in brand dialogue (Dolan et al 2015). Hence, more specifically, engagement is embodied by the ratio of likes and followers. Furthermore, dissemination through networks created by a linking of preferences—better known as the filter bubble—carries out further dissemination of information across platforms such as Facebook and Instagram (Shcherbakova and Nikiforchuk 2022). Hence, a review of comments was carried out, filtering out a number of users traced to be from outside the locality of Aretes Style. In addition to this, posts indicating the dissemination of Aretes Style's advocacy were also highlighted. These metrics were analyzed to assess the presence and impact of Aretes Style's campaigns.

Additionally, we utilized content analysis of secondary sources such as videos, brochures, and social media posts. The data from these sources shed light on how Aretes Style's actions correspond to institutional structures and efforts to promote enterprise and peace.

# 5 Results and Discussion

## 5.1 Resilience and Revival: Weaving a Future in Marawi Through Aretes Style

BARMM, established in 2019 through the Bangsamoro Organic Law, has seen a significant transformation with the rise of social enterprises, particularly in Lanao del Sur. These enterprises address socioeconomic challenges such as high poverty rates and limited access to services, contributing to sustainable peace and economic rejuvenation (UNDP, n.d.). Adducing Katayanagi and Candelaria (2023), there has been a number of businesses established following the Marawi Siege as, additionally, a measure to encourage building peace through rebuilding the community's sense of dignity and motivating coexistence. Coined through Business for Peace (B4P), a platform attributed to the UN Global Compact, examples of such businesses include Maranao Collectibles by Salika Samad, and N'ditarun Tano by Ammar Cayongcat.

Among such local enterprises include Aretes Style by Jalaloden Mustari (or more commonly known as Jal Mustari by his network), and much like the other two, this fashion-based business was established for profound reasons. In particular, Aretes Style was established to support internally displaced persons, especially women, by reviving traditional langkit weaving. This initiative symbolizes cultural pride and economic stability for Marawi (Dulay 2022), which Jal Mustari highlights as one of the more profound objectives of Aretes Style (personal communication, August 6, 2023).

Mustari envisioned Aretes Style as a platform providing sustainable employment while preserving Marawi's unique cultural heritage. By incorporating the art of langkit weaving into contemporary fashion accessories, Aretes Style counters narratives of chaos and terrorism, showcasing the strength and beauty of the Meranaw people. Each handwoven piece symbolizes peace, unity, and resilience, offering hope to a community overcoming the shadows of conflict (Dulay 2022; J. Mustari, personal communication, August 31, 2020).

Aretes Style and similar enterprises in BARMM illustrate social entrepreneurship's transformative power in conflict and recovery contexts. Mustari's vision extends beyond individual success to encompass communal healing, cultural revival, and peacebuilding. This narrative highlights entrepreneurship's potential as a vehicle for positive change and community empowerment in post-conflict scenarios. By leveraging traditional crafts and fostering economic opportunities, Aretes Style not only supports economic stability but also enhances social cohesion and cultural pride, proving crucial for long-term peace and development (J. Mustari, personal communication, August 31, 2020; UNDP 2023).

The volatile political and security landscape in BARMM has necessitated strong collaborations between Aretes Style and the local women artisans. The insecurity and instability following the Marawi Siege created a pressing need for economic opportunities and social cohesion. The urgent need to restore livelihoods and foster resilience in the community drove Aretes Style's collaboration with women weavers. Aretes Style empowered these women by giving them a platform to showcase their skills, leading to economic empowerment and the restoration of social connections and trust in the community. This partnership highlights the role of social enterprises in overcoming the difficulties that arise in post-conflict settings.

## 5.2 The Sagonsongan IDP Artisans and Tugaya Women Weavers

The Sagonsongan artisans, who were internally displaced persons (IDPs), were once vital to Marawi City's local economy. However, the Marawi Siege dramatically disrupted their community and livelihood. These individuals' relocation to the Sagonsongan site placed them in a temporary and uncertain environment. Domestic tensions often reflect the stress of their situation, highlighting the struggle to maintain hope and stability.

Amid this strife, Jal Mustari commissioned the artisans to incorporate langkit weaving into contemporary fashion, such as dresses and accessories. Crafting a new story of identity and resilience, this effort was about recovering a lost art and rebuilding a sense of self-worth and direction. With Jal Mustari's support, the Sagonsongan weavers were not just creating textiles; they were also stitching together the fabric of their collective recovery, blending endurance and optimism into every piece they created.

Examining the Tugaya Weavers' lives reveals that weaving is a deeply valued custom passed down through generations, reflecting the cultural fabric that unites families and communities. Aretes Style, under Jal Mustari's leadership, collaborates with these artisans, leveraging their traditional skills to create contemporary fashion items like woven langkit. This collaboration has led to economic opportunities such as stable market channels, enhanced product demand, and formal associations that adopt fair pricing and labor wages. These initiatives not only ensure economic viability but also foster economic independence and cultural preservation within the Meranaw community (J. Mustari, personal communication, August 31, 2020).

Before the collaboration with Aretes Style, the artisans' production was based on demand, and when orders were few, they transported their products to Marawi City's markets in the hopes of making a better profit. However, when they eventually partnered with Aretes Style, their income narratives changed. They established a stable channel to sell crafts (both orders and surplus), advertise, and consistently address their practical concerns and needs as artisans (J. Mustari, personal communication, August 31, 2020).

The artisans' journey towards empowerment reached a turning point with the formation of their association, which Aretes Style founded. This paved the way for their understanding of the benefits of having formal associations, such as adopting fair pricing and labor wages. Besides improving their organizational standing, the process of getting SEC permission in Cotabato and setting up a physical office changed the artisans' opinions of themselves, strengthening their sense of empowerment and autonomy (J. Mustari, personal communication, August 31, 2020).

Aretes Style's impact has transcended economic domains, molding the conceptions of work and income, especially in the perceptions of the younger generations. The narratives of children of artisans, elementary students, and even those as young as ten years old, who view traditional art as a leisure activity, a source of income, and a means to support their parents, shed light on the potential for economic empowerment to redefine societal and domestic roles, thereby challenging preconceived notions and fostering equality (J. Mustari, personal communication, August 31, 2020).

The empowerment of the weavers goes beyond economic development and serves as a monument to the transformative power of cooperative social enterprises in fostering strong, resilient communities. The Tugaya Weavers, with the help of Aretes Style, exemplify how a balanced combination of cultural legacy, economic empowerment, and community involvement can create a story of long-lasting positive change (J. Mustari, personal communication, August 31, 2020).

## 5.3  Empowerment Through Collaboration: The Aretes Style Partnership Model

The support and collaboration received from various organizations have played a crucial role in the growth of Aretes Style. The social enterprise received support from prominent organizations such as the United Nations Development Programme (UNDP), the United Nations Children's Fund (UNICEF), and the International Labour Organization (ILO) in alignment with the Sustainable Development Goals (SDGs), and has collaborated with The Moropreneur Inc. (TMI), Youth Co: Lab, and The Spark Project. Through such interactions providing essential support and facilitating joint activities, development for the enterprise has sailed forward, and said journey is specifically lined with amplification of influence and joint creation.

The first milestone was the YES 2 SDG Project. This project, backed by UNDP, UNICEF, and ILO, provided Aretes Style with financial support and a foundational peacebuilding and sustainable development framework through its program aimed at aggregating Bangsamoro youth sustainable development concerns and solutions. This endeavor enabled Aretes Style to transition from a visionary idea to a tangible business entity. By aligning with these prestigious organizations, Aretes Style gained significant social capital, enhancing its credibility and access to broader networks and resources.

Meanwhile, the Moropreneur Inc. (TMI) was instrumental in shaping Aretes Style's branding and mentoring aspects. TMI's collaboration provided a synergistic exchange of expertise and mission-driven strategies, enhancing Aretes Style's capacity to empower communities through social entrepreneurship. This collaboration by TMI, with the same three organizations backing the YES 2 SDG Project, offered a strategic alignment that bolstered Aretes Style's ability to navigate the complexities of operating in conflict-affected areas. It demonstrates the power of leveraging expert knowledge and networks in building a successful social enterprise (The Moropreneur 2019).

Third, Aretes Style's engagement in crowdfunding was significantly supported by Youth Co: Lab and The Spark Project. These platforms provided mentorship and guidance, helping Aretes Style connect with a broader audience and secure necessary resources for expansion. This support went beyond financial assistance; it enabled Aretes Style to maintain ownership of its projects while effectively realizing its goals. The collaborative learning and resource sharing facilitated through these platforms were crucial in building Aretes Style's social capital and enhancing its community engagement capabilities. A standout success in the enterprise's journey was its crowdfunding endeavor under The Spark Project, which notably surpassed its goal of PHP 80,000 for the "Katitinabanga project" (Mustari 2020). The success of this crowdfunding campaign opened new avenues for the enterprise, leading to its handcrafted accessories being showcased and sold through "Project Pearls," a non-profit organization based in California. This exposure was pivotal in elevating Aretes Style's brand internationally and echoing the stories and rich cultural heritage of the Meranaw people. This development is a prime example of how social capital can be effectively leveraged for business growth, meaningful storytelling, and cultural representation, reinforcing Aretes Style's commitment to social impact through entrepreneurship.

The Raheemah Economic Hub, a collaborative initiative in Marawi City funded by JICA and others, aims to revitalize local economies through skill development, especially in traditional handicrafts. It serves as a beacon of hope, empowering vulnerable sectors by enhancing capacities for economic self-sufficiency (Bagumbaran 2020). Aretes Style, a direct beneficiary, leverages this hub's training resources to upscale the abilities of local weavers and artisans. Founder Jal Mustari, capitalizing on his alumni connection with the hub's founder, navigates social norms to foster inclusion within the community, enhancing Aretes Style's social capital and enabling engagement across traditional boundaries despite being gay in a conservative society (Mustari 2020).

Moreover, Aretes Style's presence in retail stores like Ziya at Robinsons Iligan City (Aretes Style 2022) and Yours To Love Concept Store and Café in Davao City (Aretes Style 2021) demonstrates the enterprise's ability to extend its market reach and brand recognition. These partnerships provided critical market access, enabling Aretes Style to reach a wider audience and generate additional revenue streams. Such strategic collaborations are essential in building a business's brand and reputation, which are key components of social capital in social entrepreneurship.

In addition, in a collective endeavor to foster a thriving community of artisans and social entrepreneurs, together with three other Meranaw creatives, Aretes Style inaugurated a small shop on Capitol Hill, Marawi City, within the welcoming confines of the Infinitea Awar Building. This initiative serves as a vibrant hub where the artistry of Marawi's local talent, including Habibi Meranaw Delicacy, Kakaw Meranaw, AZ Craft, and Lakub, can be personally experienced and supported.

Apart from its participation in expansion and collaborative efforts, Aretes Style's active role in disaster relief and transparent communication exemplifies its commitment to community engagement and trust-building. As seen on the November 29 post of Aretes Style (2020), the enterprise's efforts in providing aid during calamities like Typhoon Ulysses and its openness in sharing these initiatives on social media reinforce its reputation and trust within the communities it serves.

While Aretes Style has benefitted from these collaborations, it is important to recognize the power dynamics involved. Aretes Style has played a receptive role in many partnerships, relying on more established organizations' resources and social capital. This dynamic presents challenges and opportunities for learning and capacity building as Aretes Style continues to grow and evolve.

In sum, Aretes Style's journey is a profound example of how strategic collaborations and the effective utilization of social capital can drive success in social entrepreneurship. Each partnership, from the foundational support of the YES 2 SDG Project to the strategic engagements with TMI, Youth Co: Lab, and The Spark Project, has uniquely contributed to Aretes Style's development. These collaborations have strengthened Aretes Style's operational capacity and enhanced its social impact, offering a model for others aspiring to influence conflict-affected communities through sustainable and socially conscious entrepreneurship.

## 5.4  Digital Transformations and Social Enterprise: Aretes Style's Journey Toward Peacebuilding, Sustainability, and Empowerment

The role of digital technology in Aretes Style highlights the power of revolutionized communication to establish social impact. This involves analyzing how the enterprise uses social media to disseminate cultural narratives, advocate for social change, market local products and services, generate income, and foster economic independence within the Meranaw community.

The story of Aretes Style continues to unfold with its strategic leverage of the power of digital channels ranging from social media pages like Facebook and Instagram to crowdfunding hosts like The Spark Project. Facebook and Instagram, both under the Meta corporation, are widely known for their unprecedented worldwide reach. These platforms each have features that add to Aretes Style's reach. These features include avenues for media postings such as photos, videos, pages, and digital feeds shaped by preference filters, allowing an opportunity to have an online presence and be discovered by a broader target audience through a snowball effect of tags. On the other hand, The Spark Project, an online crowdfunding platform, provided a more grounded support channel for Aretes Style. The Spark Project is based in the Philippines and sets an incentive-based crowdfunding model for backers and sponsors, provides a diverse category of projects, promotes community engagement, and enhances promotion and visibility for local initiatives.

To drive transformative change, championing sustainability, peacebuilding, and empowerment of marginalized peoples in post-conflict communities, Aretes Style espouses a multifaceted approach incorporating outreach efforts, engagement, economic empowerment, and stigma reduction. Discussing the tenets of the digitized aspect of their approach in detail, the power of digital transformation first touches upon the rippling outreach that extends to donors, carries over marketing knowledge, and opens opportunities for training. The key avenues and efforts shaping this outreach include social media posts that invite and encourage donors (see Fig. 1), and the opportunities created by The Spark Project's launch activity "Sparkability" and posts on the site's main page (see Fig. 2).

Aretes Style's reach is further illustrated by the steady increase in its engagement. This is not limited to the number of features on video hosting platforms such as YouTube by channels of institutions, including Youth Co: Lab and UNDP. Publicly available data, such as the number of page likes and followers (see Fig. 3), and "likes" metrics over the years provide a considerable measure of the social enterprise's exposure and the opportunities for user/audience interaction, involvement, and participation. The percentage increase for "likes", in particular, was measured using metrics data and a formula which first includes a subtraction of old likes from the new number of likes, divided by old likes, and multiplied to a hundred. Through this, there was a 16.67% increase of likes since 2020, an 11.43% increase from 2021, and a 15.38% increase from 2022. Based on these numbers, there has been a notable increase in engagement over time, suggesting a strong connection with the online audience.

Further exploring the context of economic empowerment, the gains brought on by digital transformations are first illustrated by the exceptional crowdfunding return achieved by Aretes Style via the Spark Project's site. Through the platform alone, the social enterprise was able to garner a total of 100,390 Philippine pesos, exceeding their original fund target by 25%. Second, in the same context of digital benefits, payment options such as GCash have gradually become more common, even in peripheral regions, enabling enterprises like Aretes Style to expand their business and economic networks across the country. Third, given the context provided by online crowdfunding and electronic payments, income generation has been further

**Fig. 1** Donor invitation

strengthened, cementing the relationship between Aretes Style and its partner artisans and weavers.

Consequently, the influence of these digital channels and the corresponding endeavors of the social enterprise extend beyond monetary benefits; the online presence of Aretes Style has contributed to efforts to mitigate social stigmas following the Marawi Siege. As shown in Fig. 4, an active search on Google Images results in most search returns being tagged under terms such as "Marawi Siege," "war," "battle," "rebuilding," and "Maute Group" (the terrorist group that instigated the 2017 siege). By shifting the focus to essential elements of its enterprise, such as Maranao culture, the skills and craftsmanship of the artisans now stationed in Sagonsongan, and the generations of talent carried by the Tugaya weavers, Aretes Style has emerged as a ray of hope on social media and its resulting channels and networks by combating

**Fig. 2** The spark project feature

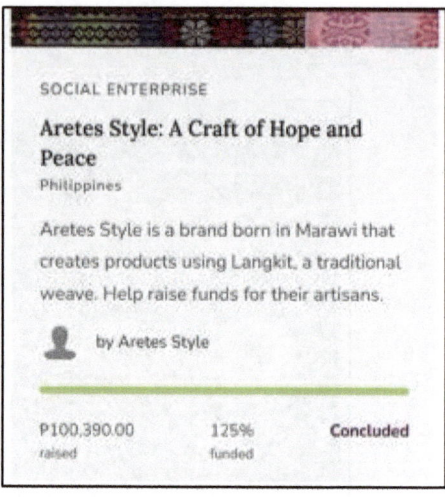

**Fig. 3** Aretes style Facebook page, current likes and following

prejudices and promoting intercultural understanding. This demonstrates the potential of social enterprises in using digital innovation and cooperative collaborations to contribute to peacebuilding and sustainability.

Subsequently, we examined the audience demographic. As highlighted by Shcherbakova and Nikiforchuk (2022) in their study on the networking capabilities of social media platforms via the linking of user preferences, the *filter bubble* shows information that aligns with the user worldview, hence the personalization of content based on user interests and behaviors are bound to link similar preferences to another, further linking users within the same interest field and behaviors. Given the varied demographic of users commenting on Aretes Style's posts, and with the page posting features by periodicals such as the *Mindanao Daily*, bubbles have then connected further outward beyond the enterprise's original locality.

**Fig. 4** 'Marawi City' Google Images Search

It is important to emphasize, however, that despite the multifaceted rewards of digital means, several noteworthy challenges arise in harnessing the full potential of said platforms. These struggles include the financial translation of engagements, the assumptions over 'reach' post features in relatively more extensive accounts in social media, and the demographic diversity of audiences. As underscored by Jal Mustari (personal communication, August 6, 2023) in his assessment of Aretes Style's digital navigation so far, features such as that of GMA news or Catriona Grey and engagements by followers of said platforms do not comprehensively address audience and revenue concerns, for online audiences may vary from creatives in the same preference networks to casual users.

## 5.5 Navigating the Institutional Landscape: Sustaining the Gains of Social Enterprises. Opportunities and Challenges

Within the socio-political and economic landscape of BARMM in general and Lanao del Sur in particular, Aretes Style stands as an exemplary social enterprise that embodies the resilience and entrepreneurial spirit of the Maranao people. This initiative is emblematic of how cultural heritage and economic development can converge within a post-conflict setting. Applying an institutionalist framework to Aretes Style's operations provides a nuanced understanding of the enterprise's influence and challenges within a complex socio-political and economic system.

The passage of the Bangsamoro Youth Act in 2020, which serves as a legislative acknowledgment of youth's pivotal role in society, has sparked youth-led endeavors. The Bangsamoro Youth Commission (BYC) has been instrumental in harnessing the vitality of the youth, directing it toward meaningful societal contributions. Aretes Style's active participation in the BYC-led Ideation Impact Challenge is a testament to its ability to tap into institutional support, fostering youth innovation in line with governance and advocating for youth empowerment. This partnership with BYC has significantly propelled Aretes Style forward, reinforcing its mission to cultivate youth leadership and social innovation within BARMM and underscoring the youth's critical impact on social evolution and community empowerment.

Aretes Style's strategic approach initially sought partnerships with government agencies like the Technical Education and Skills Development Authority (TESDA) and Masters Technological Institute of Mindanao (MTIM) for training, which, due to the pandemic, faced delays as many initiatives were put on hold. However, the tide has turned with the national government, the Bangsamoro Development Agency (BDA), and the pending enactment of the Social Enterprise Act in BARMM region, signifying a wealth of institutional support. This Act, aimed at fostering social enterprises, promises a resilient, inclusive social enterprise (RISE) for peace and sustainable development, complete with incentives and benefits.

The creation of BARMM in July 2019 marked a period of transition, not only in terms of governance structures inherited from ARMM but also in personnel, with many from the Moro Islamic Liberation Front (MILF), who were new to governance and development, taking the helm. However, the Bangsamoro Transition Authority (BTA) is committed to moral governance, openly collaborating with international development partners. This embrace of external support reflects a significant shift towards capacity building within the government and enterprise sectors.

Currently, contests initiated by BARMM that honors the peace efforts of individuals in the region like the Bangsamoro Peace Champion serve as platforms for entities like Aretes Style to amplify their voice and mission. Through such initiatives, enterprises like Aretes Style not only achieve further media exposure in the region, but are also aided with cash prizes. These developments indicate a dual evolution: the government of BARMM is finding its footing in administration and development, just as social enterprises are finding new opportunities for growth and impact. In this transitional phase, Aretes Style and similar entities are not just witnesses but active participants in shaping an ecosystem that supports innovative, socially driven business models, aligning with the region's aspirations for peace and progress.

These formal institutions, however, are juxtaposed against a political landscape marked by entrenched partisanship, which has limited the enterprise's influence, particularly in enabling political agency among its women weavers and artisans. Despite the enterprise's efforts, the political dimension presents a significant obstacle. The longstanding partisan divides within the Tugaya municipality indicate the challenges facing women weavers and artisans in exerting political agency. Support and involvement from the community have been essential to the weavers' journey. Maintaining a proactive relationship with the barangays has made pooling resources and getting more support easier. In the case of these particular weavers, because their respective barangay officials showed a seeming lack of support based on their history of supporting opposing parties during elections, they were forced to look for the aid of a different barangay to apply their initiatives formally. Aretes Style navigates a landscape where political affiliation often determines participation in civic discourse and decision-making, especially concerning economic activities that directly impact their livelihood.

Economically, the Meranaw's historical engagement in trade and entrepreneurship underscores the significance of Aretes Style's mission to rejuvenate the langkit weaving industry. However, the enterprise needs more institutional support for burgeoning startups, especially in technical knowledge and resource acquisition.

The enterprise's strategic response involves an adaptive approach to overcome infrastructural deficits, including unreliable access to electricity and internet services. Economically, Aretes Style has responded to the Marawi siege's impact on the community's commercial activity and the potential loss of the langkit weaving tradition. By addressing a heritage industry at risk, the enterprise has aligned its mission with local economic needs, filling a void left by disrupting traditional trade and business activities. The lack of formal institutional support for technical enterprise knowledge has necessitated an adaptive approach, with Aretes Style seeking knowledge externally and overcoming infrastructural challenges such as unreliable electricity and internet services.

On the other hand, Aretes Style operates within the unique socio-cultural landscape of the Maranao community, where it strives to align its business practices with local cultural values and Islamic norms. This alignment is essential in ensuring its product designs and communications resonate with the community's heritage and expectations. Mustari's leadership is pivotal in navigating the complex interplay between maintaining cultural authenticity and fostering acceptance in a society governed by traditional norms. The challenge for Aretes Style, and particularly for Jal Mustari who openly identifies as a member of the LGBTQ+ community, is further compounded by the traditional views on gender and sexual orientation prevalent in the community. Despite these potential barriers, Mustari's socio-cultural ties have been instrumental, especially with family members in politics and a progressive social enterprise founder influential in politics and Marawi society. These connections have enabled Aretes Style to gain recognition and acceptance, showcasing Mustari's adept navigation of the socio-cultural and political landscapes. The case of Aretes Style under Mustari's leadership highlights several key aspects of social entrepreneurship in culturally diverse and conservative settings: The role of government policies and socioeconomic conditions in Marawi significantly impacts the operation and success of social enterprises like Aretes Style. The alignment with cultural norms is crucial for market relevance and community acceptance. The intersectionality of Mustari's LGBTQ+ identity with traditional societal roles illustrates the complexities and challenges faced in conservative societies. Mustari's strategic use of socio-cultural and political ties underlines how informal institutions can be leveraged to navigate and overcome these challenges. The balance between adhering to traditional values and embracing modern entrepreneurial practices is critical, especially in settings where cultural authenticity and innovation must coexist.

Conversely, Aretes Style represents a strategic response to the multi-layered institutional context of BARMM, exemplifying how social enterprises can navigate complex legal, political, socio-cultural, and economic terrains. The enterprise has adapted its business model to align with the regional governance structures, socio-cultural norms, and economic realities, evidencing the transformative potential of social entrepreneurship in a context characterized by post-conflict recovery and cultural preservation.

The sustainability and long-term impact of Aretes Style's strategies are core to the institutionalist analysis. The enterprise's approach is evaluated for effectiveness and aligned with economic viability and institutional support. The focus is on the durability of Aretes Style's actions and their sustained influence on the Meranaw community and the broader BARMM region.

Aretes Style exemplifies how a social enterprise can strategically navigate and influence a multifaceted institutional landscape. Through a nuanced understanding of legal, political, socio-cultural, and economic structures, adaptive strategies, and innovative digital engagement, Aretes Style has positioned itself as a critical player in the socioeconomic empowerment and cultural preservation within BARMM. The institutionalist perspective reveals the depth of Aretes Style's integration within these systems and underscores the importance of continued agility and responsiveness to maintain its trajectory of positive social impact.

## 6  Conclusion

Aretes Style's use of technology, collaboration, and partnerships innovatively promotes peace, empowerment, and sustainability in post-conflict Marawi City. By involving the IDPs of Sagonsongan and Tugaya in producing "langkit" products, the enterprise empowers internally displaced individuals and marginalized women. This approach demonstrates how collaboration and social capital can effectively enhance economic and social resilience in conflict recovery, empowering local artisans and fostering community development.

The success of Aretes Style is largely due to its management of relationships with various networks and partners. Support from organizations like UNDP, UNICEF, and ILO has been crucial in enhancing their connections and operational capabilities. This partnership demonstrates the real-world importance of collaborations among stakeholders in improving the impact of social enterprises. This finding emphasizes the role of networks and institutional backing in expanding influence and promoting sustainable development (Maracine 2019).

Furthermore, Aretes Style has made good use of platforms such as Facebook, Instagram, and The Spark Project to achieve their goals of peacebuilding and economic empowerment. Digital tools have played a significant role in helping the enterprise promote its products, share cultural stories, and advocate for social change, expanding its influence and effectiveness. The empirical evidence based on the number of likes, shares, follower count, and commenter demographics were considered to establish Aretes Style's reach and engagement. The practical application of technology in this scenario demonstrates how social media can drive economic opportunities and societal transformation, theoretically underscoring the importance of innovation in amplifying the voices and efforts of marginalized communities in post-conflict environments.

The socio-political landscape in BARMM poses obstacles like rooted partisanship and limited institutional backing, making it hard to empower political agency among women weavers and artisans associated with the enterprise. This highlights the real-world obstacles that social enterprises face in intricate settings. It stresses the necessity for policies and inclusive governance to uplift social enterprises. This finding echoes Hidalgo et al. (2021), Larrauri and Kahl (2013), and Firchow (2018) in their argument that collaborative partnerships between social enterprises, NGOs, governments, and private sectors, fostered by supportive institutional environments, are vital for spurring innovation and amplifying impact, thereby leading to sustainable development and peacebuilding.

Upon reflection on the results, it is clear that Aretes Style's inclusive and collaborative approach has had a positive influence on the community. The enterprise's ability to leverage technology and cultivate strong partnerships has been crucial in achieving its objectives of peacebuilding and empowerment. This aligns with the body of literature discussing the role of social enterprises in post-conflict recovery, emphasizing the significance of social connections and digital advancements in promoting lasting progress. In essence, it demonstrates how social enterprises can bridge societal divides by introducing innovative and inclusive strategies.

Aretes Style serves as an example of how social enterprises can contribute to peacebuilding and economic empowerment in areas recovering from conflict through strategic partnerships and digital advancements. By tackling these obstacles and utilizing support, Aretes Style can further enhance its influence and longevity. In practice, it presents a model that can be replicated by projects worldwide, enriching conversations surrounding social entrepreneurship and sustainable development. Theoretically, it underscores the transformative capacity of enterprises in post-conflict recovery efforts, advocating for an interdisciplinary approach that is inclusive toward peacebuilding and progress.

# References

Abadie A (2004) Poverty, political freedom, and the roots of terrorism. Am Econ Rev 96(2):50–56. https://doi.org/10.1257/000282806777211847

Adjei M (2019) Women's participation in peace processes: a review of literature. J Peace Edu https://doi.org/10.1080/17400201.2019.1576515

Aldairany MA, Omar H, Quoquab F (2018) Social entrepreneurship in conflict communities. J Glob Entrep Res 8(1):361–383. https://doi.org/10.1108/JEEE-06-2017-0042

Anderlini SN (2007) Women building peace: what they do, why it matters. Lynne Rienner Publishers, Boulder

Aretes Style (2020, November 28) Today, we are officially launching Panginam scarf. [Image attached] [Status update]. Facebook. https://www.facebook.com/share/p/SShB35T7D8j5SDno/

Bagumbaran AP (2020, October 16) Eco hub, youth development center to rise in Marawi. Government of the Philippines. ReliefWeb. https://reliefweb.int/report/philippines/eco-hub-youth-development-center-rise-marawi

Bandyopadhyay S, Younas J (2011) Poverty, political freedom, and the roots of terrorism in developing countries: an empirical assessment. Econ Lett 112(2):171–175. https://doi.org/10.1016/j.econlet.2011.03.029

Banlaoi RC (ed) (2020) The Marawi Siege and its aftermath: the continuing terrorist threat. Cambridge Scholars Publishing

Barbara J (2006) Nation-building and the private sector in Afghanistan. Confl Secur Dev 6(4):559–584. https://doi.org/10.1080/14678800601066595

Battilana J, Lee M, Walker J, Dorsey C (2014) In search of the hybrid ideal. Stanford Soc Innov Rev 12(3):51–55. https://doi.org/10.48558/WF5M-8Q69

Battisti S (2019) Digital social entrepreneurs as bridges in public-private partnerships. J Soc Entrepr 10(2):135–158. https://doi.org/10.1080/19420676.2018.1541006

Bruton G, Ketchen DJ, Ireland RD (2013) Entrepreneurship as a solution to poverty. J Bus Ventur 28(6):683–689. https://doi.org/10.1016/j.jbusvent.2013.05.002

CDA Collaborative Learning Projects (2014) Business for peace: understanding and assessing corporate contributions to peace. CDA Collaborative. https://cdacollaborative.org

Colin, Hay (2002) Political analysis a critical introduction macmillan education UK London

Collier P (2007) The bottom billion: why the poorest countries are failing and what can be done about it. Oxford University Press

Cornelio J, Calamba S (2022) Going home: youth and aspirations in post-conflict Marawi, Philippines. J Youth Stud 26(5):668–685. https://doi.org/10.1080/13676261.2022.2038781

Devarapalli N, Figueira S (2015) Leveraging existing tools to help social enterprises: a case study. Procedia Eng 107:90–99. https://doi.org/10.1016/j.proeng.2015.06.093

Dolan R, Conduit J, Fahy J (2015) Social media engagement: a construct of positively and negatively valenced engagement behaviours. In: Customer engagement. Routledge, pp 102–123

Dulay P (2022, January). Aretes style: the power of community. Youth Co:Lab. YouthColab. https://www.youthcolab.org/post/aretes-style-the-power-of-community

Easterly W (2001) The elusive quest for growth: economists' adventures and misadventures in the tropics. MIT Press

Fahmy S, Hussain S (2021) War or peace tweets? The case of Pakistan. Media Int Austr 188(1):67–85. https://doi.org/10.1177/1329878X211042432

Fish A, Srinivasan R (2011) Digital labor is the new killer app. New Media Soc, 14(1):137–152. Available at: https://doi.org/10.1177/1461444811412159

Firchow P (2018) Reclaiming everyday peace: Local voices in measurement and evaluation after war. Cambridge University Press, Cambridge. https://doi.org/10.1017/9781108236140

Fu X, Avenyo E, Ghauri P (2021) Digital platforms and development: a survey of the literature. Innov Dev 11(2–3):303–321. https://doi.org/10.1080/2157930X.2021.1975361

Goodhand J (2001) Violent conflict, poverty, and chronic poverty. CPRC Working Paper No. 6. https://doi.org/10.2139/ssrn.1754535

Hidalgo G, Monticelli JM, Vargas Bortolaso I (2021) Social capital as a driver of social entrepreneurship. J Soc Entrepr 15(1):182–205. https://doi.org/10.1080/19420676.2021.1951819

Hirblinger A, Hansen J, Hoelscher K, Kolås A, Lidén K, Martins BO (2023) Digital peacebuilding: a framework for critical–reflexive engagement. Int Stud Perspect 24(3):265–284. https://doi.org/10.1093/isp/ekac015

Hoeffler A, Fearon JD (2014) Conflict and development: recent research advances and future agendas. Oxford Rev Econ Policy 30(3):596–623. https://doi.org/10.35188/UNU-WIDER/2017/404-9

Holzmann P, Gregori P (2023) The promise of digital technologies for sustainable entrepreneurship: a systematic literature review and research agenda. Int J Inf Manage 68:102593. https://doi.org/10.1016/j.ijinfomgt.2022.102593

Humphreys M, Weinstein JM (2008) Who fights? The determinants of participation in civil war. Am J Polit Sci 52(2):436–455. http://www.jstor.org/stable/25193823

International City/County Management Association (2021). Engaging marginalized communities: challenges and best practices. ICMA. https://icma.org/articles/pm-magazine/engaging-margin alized-communities-challenges-and-best-practices

Jones N, Borgman R, Ulusoy E (2015) Impact of social media on small businesses. J Small Bus Enterp Dev 22(4):611–632. https://doi.org/10.1108/JSBED-09-2013-0133

Justino P (2010) War and poverty. IDS Working Papers, 2012(391):1–29. https://doi.org/10.1111/j.2040-0209.2012.00391.x

Katayanagi M, Candelaria JL (2023, July 25) Promoting local business for Marawi's rehabilitation and peace. Fulcrum. https://fulcrum.sg/promoting-local-business-for-marawis-rehabilit ation-and-peace/

Larrauri H, Kahl A (2013) Technology for peacebuilding. Stabi: Int J Secur Dev 2(3):1–15. https://doi.org/10.5334/sta.cv

Laurell C, Sandström C, Suseno Y (2019) Assessing the interplay between crowdfunding and sustainability in social media. Technol Forecast Soc Chang 141:117–127. https://doi.org/10.1016/j.techfore.2018.12.012

Majcin J (2019) Social media challenges to peace-making and what can be done about them. Groningen J Int Law 6(2). https://doi.org/10.21827/5BF3E9C076951

Maracine L (2019) Economically empowering women as sustainable conflict resolution: a case study on building peace in Uganda through social enterprise. In: Peace, reconciliation and social justice leadership in the 21st century, building leadership bridges. Emerald Publishing Limited, Leeds, 8th edn, pp 117–130

North D (1990) Institutions, institutional change, and economic performance are all important considerations. The Cambridge University Press

Ordinario C (2017, May 28) Violence to worsen poverty in Lanao del Sur—experts. Business Mirror. https://businessmirror.com.ph/2017/05/28/violence-to-worsen-poverty-in-lanao-del-sur-experts

Paffenholz T (2016) Women in peace negotiations. In Aggestam K, Towns A (eds) Gendering diplomacy and international negotiation, studies in diplomacy and international relations. Palgrave Macmillan, pp 169–191. https://doi.org/10.1007/978-3-319-58682-3_9

Peters BG (2005) Institutional theory in political science: the new institutionalism. Ashford Colour Press, Gosport Hampshire

Porter E (2008) Why women's contribution to peacebuilding matters. Int Stud Rev 10:632–634. https://doi.org/10.1111/J.1468-2486.2008.00813.X

Ross ML (2004) What do we know about natural resources and civil war? J Peace Res 41(3):337–356. http://www.jstor.org/stable/4149748

Sachs JD (2005) The end of poverty: economic possibilities for our time. Penguin

Schmidt V (2006) Institutionalism. In Hay (ed) The state: theories and issues. Palgrave MacMillan, New York, pp 98–117

Sen A (1999) Development as freedom. Oxford University Press

Shepherd LJ (2016) Victims of violence or agents of change? Representations of women in UN peacebuilding discourse. Peacebuilding 4(2):121–135. https://doi.org/10.1080/21647259.2016.1192246

Shcherbakova O, Nikiforchuk S (2022) Social media and filter bubbles. Sci J Polonia Univ 54(5):81–88. https://doi.org/10.23856/5411

Sharifi A, Simangan D, Kaneko S (eds) (2023) Integrated approaches to peace and sustainability. World sustainability series. Springer

Sottini A, Ciambotti G (2021) Social entrepreneurship toward a sustainable peacebuilding. In: Encyclopedia of the UN sustainable development goals. Springer

Stewart F (2002) Horizontal inequalities: a neglected dimension of development. Centre for Research on Inequality, Human Security and Ethnicity. Queen Elizabeth House, University of Oxford

Tagorda Y, Magno F (2023) Unveiling the local-turn in peacebuilding: exploring the roles and contributions of Bangsamoro civil society organization in Mindanao peacebuilding. In: APISA

17th Annual Congress (APISA2023) official conference proceedings, pp 19–31. https://doi.org/10.22492/issn.2758-9374.2023.3

Thies CG (2010). Of rulers, rebels, and revenue: State capacity, civil war onset, and primary commodities. J Peace Res 47(3):321–332. http://www.jstor.org/stable/20752166

United Nations Development Programme (2023) Social entrepreneurship on human mobility in urban settings among young people in South and Southeast Asia. UNDP Philippines

United Nations Development Programme (2023, June 16) ISIP features ten social enterprises in impact showcase. UNDP. https://tinyurl.com/4w8t5bns

Väyrynen T (2010) Gender and peacebuilding. In Richmond OP (eds) Palgrave advances in peacebuilding. Palgrave Macmillan, London, pp 137–153. https://doi.org/10.1057/9780230282681_8

Weerawardena J, Sullivan Mort G (2006) Investigating social entrepreneurship: a multidimensional model. J World Bus 41(1):21–35. https://doi.org/10.1016/j.jwb.2005.09.001

World Bank (2018, March 1) Pathways for peace: inclusive approaches to preventing violent conflict. World Bank. https://www.worldbank.org/en/topic/fragilityconflictviolence/publication/pathways-for-peace-inclusive-approaches-to-preventing-violent-conflict

# Balancing the Scales: How Vietnam's Energy Transition Policies Address Equity

Ngoc Thuy Nguyen

**Abstract** Vietnam has made many commitments to energy transition, as reflected in legal documents such as the Power Development Plan VIII and National Energy Master Plan for the 2021–2030 Period With a Vision to 2050. However, no studies have evaluated whether energy-transition documents issued by the government sufficiently incorporate the equity elements. To fill in the gap, this chapter examines Vietnam's existing energy transition policies, with an emphasis on five equity aspects, including distributional justice, recognition justice, procedural justice, economic equity, and power dynamics. It also analyses the main concepts, themes, and current gaps. Out of 615 relevant documents, nine were scrutinized and demonstrated notable similarities and differences in terms of themes, concepts, and equity considerations. The chapter revealed eight recurrent focuses across all documents, including energy transition institutional frameworks and policies; climate resilience and green opportunities; human resource development; science and technology improvement; regional prosperity and inclusivity; social safety nets and well-being; communication, awareness, and community engagement; and economic diversification and revitalization. The emphasis on equity dimensions was limited in all publications. Further efforts are imperative to ensure the policy framework reflects a commitment to support the affected communities and enhance public engagement in the energy transition process. While improvements on institutional policies and technical solutions have been addressed more comprehensively in the Vietnamese legal documents, more attention should be brought to the socio-economic burdens affecting certain communities. Connecting social priorities with energy transition is crucial to create effective policies and prompt actions that benefit different social groups.

**Keywords** Energy sector · Equity · Vietnam · Policies · Just energy transition

N. T. Nguyen (✉)
Centre for Environmental Policy, Imperial College London, London SW7 2BX, UK
e-mail: n.nguyen21@imperial.ac.uk

© The Author(s), under exclusive license to Springer Nature Singapore Pte Ltd. 2024
A. Sharifi et al. (eds.), *Navigating Peace and Sustainability in an Increasingly Complex World*, World Sustainability Series, https://doi.org/10.1007/978-981-97-8772-2_5

# 1 Introduction

In the contemporary landscape, attaining a just energy transition has become a pronounced priority, with countries grappling with the challenges of mitigating climate change effects while ensuring energy security. Amidst this context, Vietnam, a rapidly developing country in Southeast Asia, finds itself at a pivotal juncture in the transition of its energy system due to a convergence of factors. The nation has a burgeoning economy, rapid industrialization, and urbanization, with its annual GDP growth rate demonstrating an upward trend. Despite crises, Vietnam's GDP growth rate was eight percent in 2022 and is expected to remain at 6.3 and 6.5% in 2023 and 2024, respectively (World Bank 2023). Consequently, the nation's energy demand has witnessed a constant increase. After 2022, the electricity demand in Vietnam is projected to rise at a rate of eight percent (Fitch Ratings 2022), prompting various issues regarding resource accessibility and environmental deterioration.

Additionally, geographical features such as an extensive coastline spanning over 3000 km, river deltas, and complex terrain render Vietnam acutely susceptible to natural disasters and climate change effects (World Bank Group and Asian Development Bank 2021). Over the past two decades, natural disasters have cost Vietnam roughly $250 million (Cong et al. 2016). Extreme weather events and disasters also affect approximately one million people per year, particularly those whose primary means of livelihood hinge on natural resources and agriculture (Nguyen et al. 2019). Vietnam's vulnerability to climate change impacts, environmental concerns, and substantial surge in energy demand due to economic growth, industrialization, and urbanization underscore the urgency for a strategic shift from fossil fuels, which constitute a significant portion of energy needs, toward low-carbon energy sources such as wind and solar to ensure economic stability and energy security.

This shift necessitates meticulous planning in Vietnam's directions and actions regarding energy generation, distribution, and consumption. The Resolution 55-NQ/TW dated February 11, 2020, by the Politburo, is one of the crucial documents that set out the strategic direction for national energy development. The Resolution emphasizes the importance of prioritizing fast and sustainable energy development while ensuring favorable conditions for different economic sectors. Following the implementation of Resolution 55, the government has shown strong awareness and determination to transition the energy sector through various initiatives, including the Just Energy Transition Partnership. Key issues related to the energy sector, such as energy supply security, price rationality and competitiveness, greenhouse gas emissions, and sustainability, have been thoroughly discussed and addressed at the national level. These efforts are reflected in strategic plans for the energy sector, such as the Power Development Plan (PDP) VIII. Besides these, effective energy transition also necessitates people-centered solutions that ensure the equitable distribution of benefits and disadvantages among its diverse population. Since the nation's socioeconomic milieu is characterized by extensive diversity, entailing varying income strata, inequities in access to resources, and urban and rural settings, discerning the ramifications of energy transition policies across these contexts and understanding the nuances of the socio-economic heterogeneity are pivotal.

Delving into the equity dimension within Vietnam's energy transition, this chapter investigates the interconnections between energy transition policies and equity. It aims to answer the question: "How can energy transition policies in Vietnam be bolstered to give precedence to equity?" The chapter is guided by four key objectives: conducting an in-depth analysis of the current energy transition policies in Vietnam with a focus on their main objectives, design, and mechanisms; evaluating how equity considerations are incorporated into these existing policies; understanding the level of involvement that different stakeholders have in contributing to these policies; and proposing recommendations on how Vietnam's energy transition policies could align more closely with principles of equity.

The subsequent section provides an overview of Vietnam's energy landscape, the need and challenges for energy transition, and the contribution of this work. Thereafter, the chapter explores different theoretical groundings and presents a conceptual framework articulating a structured approach to developing and implementing just energy transition policies. Using this framework, the chapter assesses the energy transition policies in Vietnam, focusing primarily on the extent to which the equity aspect has been incorporated and the levels of engagement by different stakeholders. The final section outlines policy recommendations that prioritize equity to ensure an inclusive energy transition in Vietnam.

## 1.1 Overview of Vietnam's Energy Transition and Importance of Equity Considerations

Vietnam's historical energy paths have been marked by a dependence on fossil fuels such as coal and natural gas to power its economic growth (Fig. 1). The coal share in electricity generation has steadily increased over the years, reaching 38% in 2022 (Ember, n.d.). Given the escalating concerns over climate change and the global commitment to renewable energy sources, these paths are no longer tenable. The need to reduce greenhouse gas emissions and transition toward net zero has propelled Vietnam to adopt progressive energy transition measures. As a signatory to the Global Coal to Clean Power Transition Statement, Vietnam is committed to transitioning away from unabated coal power generation by the 2040s (Vietnamplus 2023). A steady stream of political decisions and guidelines on energy-related matters further demonstrates the nation's goal of achieving net zero and transforming its energy system. This includes the Decision for Implementing Vietnam's Commitments at COP26 (Decision No. 2157/QD-TTg) and the recently adopted Power Development Plan VIII (PDP VIII), which establishes the overarching objectives and strategies for the national power sector.

In parallel with its commitment to a sustainable energy transition, Vietnam faces the challenge of energy equity. Transitioning from fossil fuels is a multifaceted

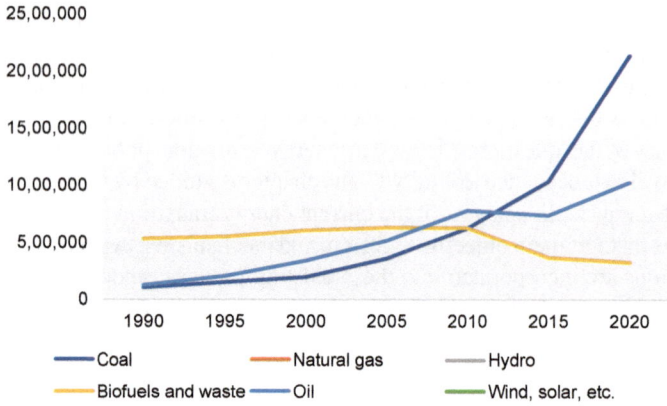

**Fig. 1** The total energy supply by source in Vietnam from 1990 to 2020. *Source* IEA

process that requires a comprehensive and well-coordinated plan. This plan must address not only environmental concerns and the advancement towards renewable energy sources but also the negative socio-economic effects on workers impacted by the closure of coal power plants, 82% of which are less than ten years old (Nguyen 2023), and the communities heavily reliant on coal-related activities. One immediate consequence of the transition is the displacement of workers in fossil fuel sectors, such as coal mining, oil extraction, and natural gas production. In Vietnam, approximately 100,000 people employed in fossil fuel extraction or coal-fired power plants are likely to experience severe economic disruption, including significant wage loss due to the decline in job opportunities and difficulties in finding new ones because of a lack of necessary skills. Within this group, 78,000 workers in the coal mining sector face an annual reduction of approximately three percent (Bao Phap Luat 2023).

Transitioning the energy sector is interconnected with various policy domains and impacts not only the environment and economy but also society. To achieve its energy goals, Vietnam must not focus solely on a single policy area or technical domain. Instead, it needs to address inherent socio-economic disparities and ensure the equitable distribution of benefits associated with the energy transition among its population. Reaching this equilibrium requires a nuanced understanding and incorporation of equity considerations across all stages of energy transition-related policy development, implementation, and impact assessment (Barlow et al. 2022). Equity is a crucial facet that cannot be underestimated, as differential access to clean energy resources might exacerbate inherent socio-economic and environmental issues. From a social justice viewpoint, marginalized areas tend to shoulder a disproportionate environmental burden and compromised health outcomes due to heavy dependence on conventional energy sources (Barlow, Tapio, and Tarekegne 2022; Chen et al. 2022). Hence, it is crucial to create opportunities for people and communities, particularly those overlooked or bypassed by industry-led policies, to access sufficient, affordable, reliable, and renewable energy of their choice.

In the context of Vietnam, the concept of a just energy transition aligns with the green growth agenda, which considers economic growth, environmental protection, and human development to be mutually reinforcing. However, one of the shortcomings of the green growth agenda is that it has yet to address various patterns and forms of inequities, including procedural justice, distributive justice, and recognition justice. There is a growing body of literature on the energy sector in Vietnam, but most of it does not emphasize the diverse equity aspects of energy transition policies. As Vietnam has progressively signed international agreements to facilitate a just energy transition, it is crucial to understand how just considerations are addressed in the current regulatory and legal frameworks. This chapter aims to address these current gaps.

## 2 Conceptual Framework for Equitable Energy Transition Policies

Energy transition is not solely an imperative due to environmental concerns but also a socio-economic undertaking. Ensuring the "just" element of the transition requires a comprehensive framework that blends theoretical foundations with empirical strategies. This chapter analyzes various principles related to environmental justice, adaptive governance, and political ecology to develop a framework that offers a structured approach for creating and implementing policies aimed at ensuring an equitable energy transition.

### 2.1 Environmental Justice Principles

The concept of environmental justice emphasizes the importance of equitable distribution of environmental benefits and burdens to achieve sustainability (Schlosberg and Collins 2014). Energy transition policies should, therefore, be grounded in the principles of environmental justice, ensuring that no demographic groups disproportionately shoulder the costs of or are excluded from the benefits of the energy transition (Jenkins et al. 2016; Sovacool and Dworkin 2015). Environmental justice is multidimensional and has been interpreted in various ways. This chapter perceives environmental justice as encompassing three key facets: distributive, participatory, and recognition justice.

Distributive justice refers to the equitable distribution of resources and environmental benefits. In the context of energy transition, this concept guides the identification of vulnerable and underserved populations (Cook and Hegtvedt 1983; Lamont 2017). A comprehensive socioeconomic analysis and mapping can reveal the socio-economic groups that face high levels of risk related to energy poverty, job displacement, and environmental burdens.

Participatory justice suggests that different stakeholders, including communities, should have the right to participate in decision-making processes (Stephens 1986). Meaningful engagement with local communities, particularly those affected by the energy transition, is crucial to ensure that their needs, concerns, and voices are accurately reflected in policy formulation and implementation.

Recognition justice asserts that the rights, livelihood priorities, preferences, and knowledge of different socioeconomic groups, particularly the vulnerable or high-risk ones due to the energy transition, should be acknowledged. Recognition is not only a fundamental goal for justice but also a foundation for achieving distributive and participatory justice (Honneth 2004; Schweiger 2019). Without understanding and incorporating the insights from those most acutely affected by the energy transition, there is a risk of overlooking contextually appropriate strategies to address issues concerning the energy system.

## 2.2   Political Ecology

The theoretical foundation of political ecology emphasizes the socio-political facets of environmental matters, focusing on the power dynamics that dictate the access to and distribution of resources. Policies should reflect the core tenet of political ecology, demonstrating a keen recognition of the socio-political dimensions of energy issues, existing power imbalances, and a willingness to rectify these disparities (Ahlborg and Nightingale 2018; Robertson 2015; Robbins 2019; Sovacool 2021).

Another key aspect of political ecology is environmental governance, which entails a detailed examination of the governance systems and institutions responsible for regulating and managing environmental and energy-related issues (Ahlborg and Nightingale 2018). This approach reveals structural shortcomings and other areas requiring improvement, thereby contributing to a participatory approach to decision-making (Robbins 2019). Environmental governance is particularly relevant to Vietnam, as it directly impacts the formulation and implementation of energy transition-related policies. Decision-making processes should be inclusive and transparent, allowing diverse stakeholders to contribute meaningfully.

Political ecology further emphasizes the importance of community empowerment and engagement in decision-making processes, ensuring that policies are attuned to the needs and values of the population (Bixler et al. 2015; Osborne et al. 2021). This aligns closely with the concept of recognition justice, which involves acknowledging the different identities and rights of the populations mentioned previously.

The political economy of energy constitutes an integral part of political ecology. It illuminates the complex interrelationships between political, economic, social, and environmental aspects in resource management and provides insights into how economic variables and forces such as profit motives, corporate interests, and investment strategies shape energy policies and practices (Robbins 2019; Swyngedouw 2009). For instance, economic interests might favor one mode of energy production or distribution over another depending on resource availability or profitability. However, this tendency might inadvertently sideline those reliant on alternative energy sources or living in regions with limited access to those energy sources, leading to greater energy burdens or environmental repercussions for these groups.

## 2.3  Conceptual Framework

Figure 2 illustrates the conceptual framework that integrates the notions of environmental justice and political ecology as described in Sects. 3.1 and 3.2. The confluence of these concepts provides a nuanced and comprehensive approach to understanding and addressing the challenges and opportunities presented by the energy transition.

Political ecology, rooted in the critical exploration of power dynamics and the interplay of socio-economic conditions and the policy environment, provides a foundation for examining the socio-economic-political intricacies inherent in environmental and energy-related issues. This framework combines various aspects and implications of political ecology, demonstrating how understanding social and cultural norms and the complexities of power dynamics is crucial to unveiling how diverse actors—including states, local communities, and businesses—impact and are impacted by the energy transition. It also elucidates the intricate linkages between different aspects of the energy transition and the broader matrices of societal relationships, organizations, and values.

The integration of environmental justice principles into this framework enhances the analytical paradigm with equity dimensions. Environmental justice emphasizes

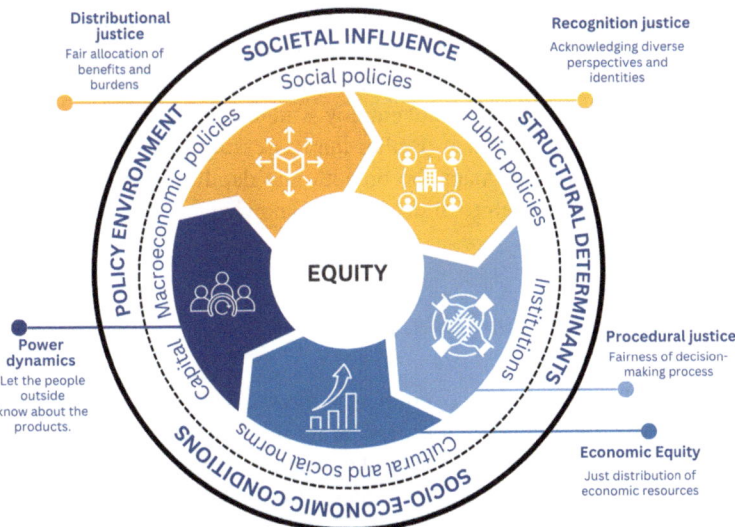

**Fig. 2** Conceptual framework linking political ecology and environmental justice dimensions

the importance of equitably distributing environmental benefits and burdens and mitigating the disparate effects of public, macroeconomic, and social policies on different
communities. Hence, the framework is interdisciplinary, considering a comprehensive, "whole-systems" spectrum of environmental, social, economic, and political
dimensions simultaneously. It illustrates that energy and environmental issues are
interlinked with socio-political-economic structures and that resolving them requires
a holistic understanding of justice concerns, power dynamics, and environmental
contexts.

By incorporating these equity dimensions, this chapter underscores that the
concept of a just energy transition extends beyond employment to encompass the
equitable distribution of costs and benefits, the allocation of risks, energy vulnerabilities, and institutional arrangements. This framework highlights the underlying
causes of inequities emerging from interactions within the socio-economic-technical
system.

## 3   Methods: Scoping Review

### 3.1   Purpose

The chapter employs a scoping review method, a systematic knowledge synthesis
approach designed to provide insights into the pre-determined overarching question and follows the Preferred Reporting Items for Systematic Reviews and Meta-
Analyses Extension for Scoping Reviews (PRISMA-ScR) guidelines. This method
explores and maps the content of national energy transition policies and practices
in Vietnam, with a focus on equity aspects. It offers a thorough overview of the key
concepts, research gaps, and thematic areas related to the topic. Unlike a systematic review, a scoping review does not employ stringent inclusion criteria but adopts
a broader, more general approach. Rather than assessing the quality of individual
studies, this scoping review maps the breadth and depth of available evidence to
provide a comprehensive picture of the existing knowledge base.

### 3.2   Search Strategy and Data Sources

This chapter utilized Vietnam's government websites, which contain all Vietnamese
policies, strategies, and interventions, including those aimed at energy transitions.
The search terms were developed using keywords covering key concepts such as
"energy," "transition," "equity," and "Vietnam." Boolean operators (AND, OR) were
applied to refine the search queries and ensure the inclusion of relevant publications.

## 3.3   Inclusion and Exclusion Criteria

The eligibility criteria for this study were broad to ensure that a wide array of legal documents, journals, and reports focusing on energy transition were captured. The specific inclusion and exclusion criteria are shown in Table 1.

The decision to narrow the search to include publications addressing both the energy transition and equity aspects stemmed from the objective of understanding the intersection between these two spheres. By focusing on publications that address these intertwined concepts, the chapter aims to reveal the intricacies of how equity considerations are incorporated into the energy transition process. Expanding the search to other documents related to the energy sector might dilute the analytical depth of these aspects. The current scope facilitates a focused, empirically pertinent, and rigorous examination of the interconnection between energy transition and equity.

Additionally, the chapter confined its scope to publications issued after 2021, as that year marked a significant milestone in the environmental policy landscape of Vietnam. During the 26th UN Climate Change Conference of the Parties (COP26), Vietnam's Prime Minister, Pham Minh Chinh, committed to attaining net-zero emissions. This commitment signifies a substantial shift toward a low-carbon and sustainable trajectory. Consequently, this chapter emphasizes post-2021 publications to capture the evolving discourse and policy paradigms in preparation for or response to this event. This approach aims to provide contemporary and contextually pertinent insights into Vietnam's energy transition efforts. Furthermore, the chapter reviews the most recent versions of documents updated based on their previous iterations.

**Table 1**  Inclusion and exclusion criteria for the review

|  | Inclusion criteria | Exclusion criteria |
|---|---|---|
| Type of publication | All publications (e.g., peer-reviewed journals, conference proceedings, reports, legal documents, grey literature) relevant to energy transition policies, equity, and their interlinkage in the Vietnamese context | Publications of which the primary focus is not on energy transition policies, equity, and their interlinkage in Vietnam |
| Type of study | Implemented legal and regulatory documents (e.g., decisions, decrees, plans) | Not applicable |
| Language | English and Vietnamese | Non-English and non-Vietnamese |
| Time frame | 2021 onward | Before 2021 |

### 3.4 Data Selection and Extraction

This chapter adopts a two-stage screening process. In the first stage, the researcher screened the titles and abstracts of the identified publications to ensure their relevance to the research topic and question. Publications passing the first stage were then subjected to a full-text review in the second stage. Those selected after this comprehensive review were included in the scoping review. The researcher extracted data into components such as author(s), publication year, study type, objectives, main findings, and key concepts related to energy transition policies and equity. The data was managed using Excel.

### 3.5 Data Synthesis and Analysis

The data synthesis and analysis process involved thematic analysis to pinpoint common themes and patterns within the selected publications. The publications were systematically coded and categorized into principal themes underpinning policy interventions and strategies. This chapter employed a combination of deductive and inductive thematic analysis techniques to identify, develop, and evaluate the emerging themes. For deductive analysis, the conceptual framework (Fig. 2), which includes concepts such as political ecology and environmental justice, guided the initial coding process. Deductive analysis served as the starting point for organizing and categorizing the data. In addition to deductive analysis, inductive analysis was used to explore new, unexpected themes that emerged directly from the data itself. This dual approach ensured a comprehensive examination of the data, incorporating both existing knowledge and novel insights generated from the chapter's findings. The researcher re-read the full-text documents multiple times to ensure all relevant themes were captured. This process laid the foundation for a comparative analysis across publications.

After identifying themes and equity concerns, the researcher developed a coding scheme, assigning a specific code to each theme. This scheme facilitated the systematic categorization of instances when a theme appeared within the text, and a tally of the occurrences of each theme was kept. Subsequently, the frequencies of these themes were calculated, either manually or with the aid of qualitative data analysis software. The resulting sums indicated how often each theme appeared in the text. The researcher then proceeded to rate these frequencies, classifying themes into high, medium, or low frequency categories based on their occurrence. To rate these frequencies, thresholds were established based on the distribution of counts. Additionally, the researcher considered the significance of each theme, defined as the extent to which the presence of the theme enriched the overall content.

# 4 Results

Using keywords including "energy," "transition," and "equity," the search yielded a total of 615 documents. After the screening processes, nine documents were chosen for in-depth investigation as they were most pertinent to this research. All of them were decisions issued by different authorities, including the Prime Minister (n = 7) based on proposals by the ministries, the State Bank of Vietnam (n = 1) based on proposals by its associated department, and the People's Committee of Hung Yen (n = 1). Eight documents were at the national level, while one was at the provincial level. Annex 1 provides a systematic map that offers a comprehensive overview of the included publications. Table 2 provides a brief description of the included publications.

## 4.1 Key Themes of the Publications

This chapter identified eight main themes across all documents. These are described in Table 3.

All the nine included documents addressed all the themes, but there are slight differences in focus across the documents. The heat map below demonstrates the theme coverage by each included document (Fig. 3).

The documents encompass various facets of regulatory mechanisms, legal frameworks, and governmental agencies responsible for navigating and driving the energy transition in Vietnam. They collectively highlight the government's commitment to facilitating the nation's transition to sustainable growth and outline the roles, responsibilities, and specific tasks of different governmental entities. Each document also includes a proposed timeline for these tasks. However, only three decisions place a heavy focus on the governance of the energy sector and provide sector-specific targets and metrics. The remaining six articulate the overarching vision of driving Vietnam's sustainable growth but do not outline clear targets or action plans within the energy sector. In these six documents, the energy transition is considered a subsidiary theme within the broader sustainability and green growth objectives.

Additionally, all documents emphasize the importance of coordination across different policy levels and governmental entities, identifying the leading and coordinating entities for each task. However, they do not detail specific implementation or coordination mechanisms, creating ambiguity regarding responsibilities and potentially leading to overlapping tasks.

All nine documents highlighted the crucial importance of preparedness and transition for climate change mitigation, discussing green economic growth with an emphasis on sectors like renewable energy. Five decisions strongly focused on advancing Vietnam's green economic opportunities within the landscape of climate resilience, while four were geared towards reducing greenhouse gas emissions to cope with the changing climate. One provincial decision focused on the development of

**Table 2** Brief details of the included publications

| Title | Year | Type | Policy level | Issuing authority | Main responsible authority | Legitimacy and legal compliance |
|---|---|---|---|---|---|---|
| **Prime Minister's Decision 888/QD-TTg** on approval for scheme setting out tasks and solutions for implementation of outcomes of the 26th Conference of Parties to the United Nations Framework Convention on Climate Change | 2022 | Decision | National | Prime Minister | Ministry of Natural Resources and Environment | – 2015 Law on Governmental Organization<br>– Law on amendments to the Law on Governmental Organization and the Law on Local Government Organization dated November 22, 2019<br>– Resolution No. 01/NQ-CP dated January 08, 2022 |
| **Prime Minister's Decision 876/QD-TTg** on approving the action program for transition to green energy and mitigation of carbon dioxide and methane emissions from transportation | 2022 | Decision | National | Prime Minister | Ministry of Transportation | – 2015 Law on Governmental Organization<br>– Law on amendments to the Law on Governmental Organization and the Law on Local Government Organization dated November 22, 2019 |

(continued)

**Table 2** (continued)

| Title | Year | Type | Policy level | Issuing authority | Main responsible authority | Legitimacy and legal compliance |
|---|---|---|---|---|---|---|
| **Decision 500/QD-TTg** on approving the National Power Development Plan for the 2021–2030 period, with a vision to 2050 | 2023 | Decision | National | Prime Minister | Ministry of Industry and Trade | – 2015 Law on Governmental Organization<br>– Law on amendments to the Law on Governmental Organization and the Law on Local Government Organization dated November 22, 2019<br>– Law on Planning dated November 24, 2017; Law on Electricity dated December 3, 2004<br>– Law on amendment to the Law on Electricity dated November 20, 2012<br>– Resolution No. 61/2022/QH15 dated June 16, 2022 of the National Assembly<br>– Resolution No. 81/2023/QH15 dated January 9, 2023 of the National Assembly |

(continued)

**Table 2** (continued)

| Title | Year | Type | Policy level | Issuing authority | Main responsible authority | Legitimacy and legal compliance |
|---|---|---|---|---|---|---|
| **Decision 893/QD-TTg** on approval of the National Energy Master Plan for the period 2021–2030, with a vision to 2050 | 2023 | Decision | National | Prime Minister | Ministry of Industry and Trade | – 2015 Law on Governmental Organization <br> – Law on amendments to the Law on Governmental Organization and the Law on Local Government Organization dated November 22, 2019 <br> – Planning Law dated November 24, 2017; Resolution No. 61/2022/QH15 dated June 16, 2022 of the National Assembly <br> – Resolution No. 81/2023/QH15 dated January 9, 2023 of the National Assembly <br> – Decree No. 37/2019/ND-CP dated May 7, 2019 <br> – Report No. 4225/TTr-BCT dated July 3, 2023; … |

(continued)

**Table 2** (continued)

| Title | Year | Type | Policy level | Issuing authority | Main responsible authority | Legitimacy and legal compliance |
|---|---|---|---|---|---|---|
| **Decision 1009/QD-TTg** on Approval of the implementation scheme for the political declaration to establish a just energy transition partnership | 2023 | Decision | National | Prime Minister | Ministry of Natural Resources and Environment | – 2015 Law on Governmental Organization<br>– Law on amendments to the Law on Governmental Organization and the Law on Local Government Organization dated November 22, 2019<br>– Political Declaration on establishing a fair energy transition partnership with Vietnam … |

(continued)

**Table 2** (continued)

| Title | Year | Type | Policy level | Issuing authority | Main responsible authority | Legitimacy and legal compliance |
|---|---|---|---|---|---|---|
| **Decision 1408/QD-NHNN** on the issuance of the banking sector's action plan to implement the National Strategy on Green Growth Period 2021–2030 and project on tasks and solutions for implementing the results of the 26th conference of parties UN Framework Convention on Climate Change | 2023 | Decision | National | State Bank Governor | State Bank of Vietnam | – Law on the State Bank of Vietnam dated June 16, 2010<br>– Law on Credit Institutions dated June 16, 2010<br>– Law amending and supplementing a number of articles of the Law on Credit Institutions dated November 20, 2017<br>– Decree No. 102/2022/ND-CP dated December 12, 2022<br>– Decision No. 1658/QD-TTg dated October 1<br>– Decision No. 882/QD-TTg dated July 22, 2022<br>– Decision No. 888/QD-TTg dated July 25, 2022<br>– Decision No. 896/QD-TTg dated July 26, 2022 of the Prime Minister<br>– Decision No. 687/QD-TTg dated June 7, 2022<br>– Decision No. 889/QD-TTg dated June 24, 2020 |

(continued)

**Table 2** (continued)

| Title | Year | Type | Policy level | Issuing authority | Main responsible authority | Legitimacy and legal compliance |
|---|---|---|---|---|---|---|
| **Decision 896/QD-TTg** on approval of the national climate change strategy for the period up to 2050 | 2023 | Decision | National | Prime Minister | Ministry of Natural Resources and Environment | – 2015 Law on Governmental Organization<br>– Law on amendments to the Law on Governmental Organization and the Law on Local Government Organization dated November 22, 2019<br>– Resolution No. 06/NQ-CP dated January 21, 2021<br>– Conclusion No. 56-KL/TW dated August 23, 2019 of the Politburo<br>– Resolution No. 50/NQ-CP dated May 20, 2021 |

(continued)

**Table 2** (continued)

| Title | Year | Type | Policy level | Issuing authority | Main responsible authority | Legitimacy and legal compliance |
|---|---|---|---|---|---|---|
| **Decision 1658/QD-TTg** on approval of the National Green Growth Strategy for Period 2021–2030, with a vision to 2050 | 2021 | Decision | National | Prime Minister | Ministry of Planning and Investment | – 2015 Law on Governmental Organization<br>– Law on amendments to the Law on Governmental Organization and the Law on Local Government Organization dated November 22, 2019<br>– Resolution No. 24-NQ/TW dated June 3, 2013<br>– Conclusion No. 56-KL/TW dated August 23, 2019 of the Politburo |

(continued)

**Table 2** (continued)

| Title | Year | Type | Policy level | Issuing authority | Main responsible authority | Legitimacy and legal compliance |
|---|---|---|---|---|---|---|
| **Decision 1506/QD-UBND** on approval of the project for development of roof solar power in the province until 2030, with a vision to 2045 | 2023 | Decision | Provincial | People's Committee of Hung Yen Province | People's Committee of Hung Yen Province | – 2015 Law on Governmental Organization<br>– Law on amendments to the Law on Governmental Organization and the Law on Local Government Organization dated November 22, 2019<br>– Electricity Law dated December 3, 2004<br>– Law amending and supplementing a number of articles of the Electricity Law dated November 20, 2012<br>– Resolution No. 55-NQ/TW dated February 11, 2020 of the Politburo<br>– Decision No. 500/QD-TTg dated May 15, 2023 of the Prime Minister<br>– Plan No. 251-KH/TU dated July 29, 2020 of the Provincial Party Committee<br>– Plan No. 119/KH-UBND dated September 22, 2020 of the Provincial People's Committee<br>– Decision No. 1014/QD-UBND dated May 6, 2022 of the Provincial People's Committee |

**Table 3** Recurrent themes across the studied documents and their associated description

| Theme | Description |
|---|---|
| Energy transition institutional frameworks and policies | This theme focusses on the availability and/ or proposed improvement of the current institutional structures, regulations, policies established by governments to facilitate energy transition |
| Climate resilience and green opportunities | This theme includes strategies and initiatives aiming to enhance climate resilience and green economic opportunities |
| Human resource development | This theme focusses on efforts to equip the workforce, particularly displaced workers, with the necessary skills and knowledge to adapt to the changing job landscape due to energy transition |
| Science and technology improvement | The theme pertains to investments, initiatives, and other efforts concerning scientific research and technological advancements propelling innovation in the energy sector. These might include improvement in the energy efficiency of energy technologies/ industries, and so on |
| Regional prosperity and inclusivity | This theme refers to initiatives, strategies, and/ or proposals to improve the economy and inclusivity across regions, especially those negatively impacted by energy transition |
| Social safety nets and well-being | This theme refers to initiatives, strategies, and/ or proposals that offer social support and ensure wellbeing of the communities affected during the period of energy transition. Particularly, there is a focus on social support for displaced workers and communities |
| Communication, awareness, and community engagement | This theme highlights initiatives, strategies, and/ or proposals to promote communication, awareness |
| Economic diversification and revitalisation | This theme focusses on opening economic opportunities to replace declining industries or update existing ones. It demonstrates a commitment to entrepreneurship and modernisation of regional economies |

rooftop solar panels without linking it to climate resilience. However, overall, the synergies between climate adaptation, mitigation, and energy transition were not clearly delineated within the broader sustainability objectives.

The theme of human resource development was featured in all documents, though the level of coverage was moderate. All documents emphasized the importance of workforce training programs, highlighting the necessity to upskill and reskill workers

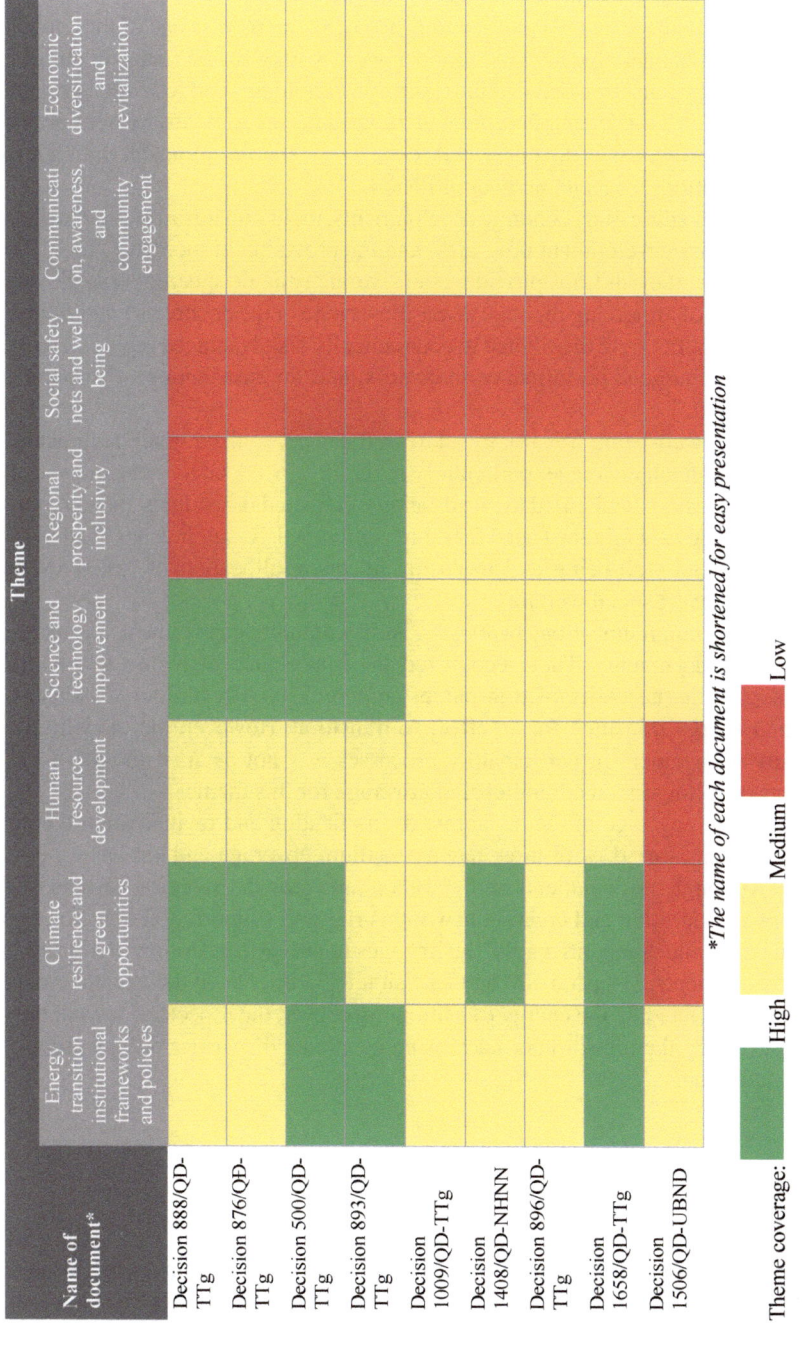

*The name of each document is shortened for easy presentation

**Fig. 3** Heat map showing the extent to which each document covers the overarching themes

for emerging industries. However, they did not detail specific strategies, topics, or programs, nor did they provide a roadmap of the required resources.

The theme of science and technology improvement emerged as one of the central facets of the documents. Specifically, four decisions with a strong focus on climate change and energy transition detailed the types of technologies and research required for the transition. All documents accentuated the need for investments in research and development. However, they did not provide details on technology transfer initiatives, practical applications, or technological diffusion.

All documents addressed economic developments, including infrastructure investments, community development programs, and the provision of incentives for businesses. However, they did not provide extensive analysis or specific strategies to achieve the goal of mitigating regional disparities. None of the documents mentioned ensuring that benefits were distributed to economically disadvantaged regions. Additionally, specific targets, performance indicators, and the distribution of resources were not detailed.

While social well-being was briefly addressed and recognized in all documents, there was no in-depth coverage, particularly regarding social safety nets and social support mechanisms to safeguard the well-being of affected individuals and communities. This implies a heightened need for more comprehensive approaches to address the issue of social well-being and to ensure adequate allocation of resources to enhance and support social welfare.

The theme of communication, awareness, and community engagement was recurrent across all documents. They recognized the importance of public awareness campaigns and the necessity of transparent communication regarding sustainable growth and energy transition for an effective transition. However, the modality of community engagement or participatory processes was not detailed in any of the documents, resulting in a medium level of coverage for this theme.

The depth of coverage of the economic diversification and revitalization theme across documents varied, with three having medium coverage and six having low coverage. Although the documents outlined economic growth strategies, particularly in terms of ameliorating and creating new industries and supporting innovation for improved economic competitiveness, the linkages to energy transition were limited. Only two documents, Decision 500 and Decision 893, articulated the need for revitalizing areas affected by the energy transition. Moreover, the aspects of revitalizing areas impacted by the transition or addressing economic disparities did not receive adequate attention.

## 4.2   Aspects of Energy Equity

The selected documents recognize five justice dimensions: distributional justice, procedural justice, recognition justice, economic equity, and power dynamics to varying degrees. While these justice dimensions were acknowledged across the documents, none demonstrated a high level of coverage. The heat map below shows the coverage of equity aspects by each document (Fig. 4).

Only two documents had a medium coverage of distributional justice, while the rest exhibited significant gaps and did not specify the areas that needed more support during the energy transition. It is important that access to resources and opportunities are ensured for all social groups within the society and they have the chance to contribute and benefit from the energy transition.

The aspect of recognition justice was not covered in all documents. They did not highlight the need to ensure representation by different stakeholders, particularly those affected by the transition, in institutions or decision-making processes. However, there was recognition of the need to collaborate with and involve international organizations.

Procedural justice was a recurrent topic across all documents, with an emphasis on the cruciality of effective governance and the participation of different stakeholders.

| Name of document* | Aspects of energy equity | | | | |
|---|---|---|---|---|---|
| | Distributional justice | Recognition justice | Procedural justice | Economic equity | Power dynamics |
| Decision 888/QD-TTg | | | | | |
| Decision 876/QĐ-TTg | | | | | |
| Decision 500/QD-TTg | | | | | |
| Decision 893/QD-TTg | | | | | |
| Decision 1009/QD-TTg | | | | | |
| Decision 1408/QD-NHNN | | | | | |
| Decision 896/QD-TTg | | | | | |
| Decision 1658/QD-TTg | | | | | |
| Decision 1506/QD-UBND | | | | | |

*The name of each document is shortened for easy presentation*

Level of coverage:  Medium   Low

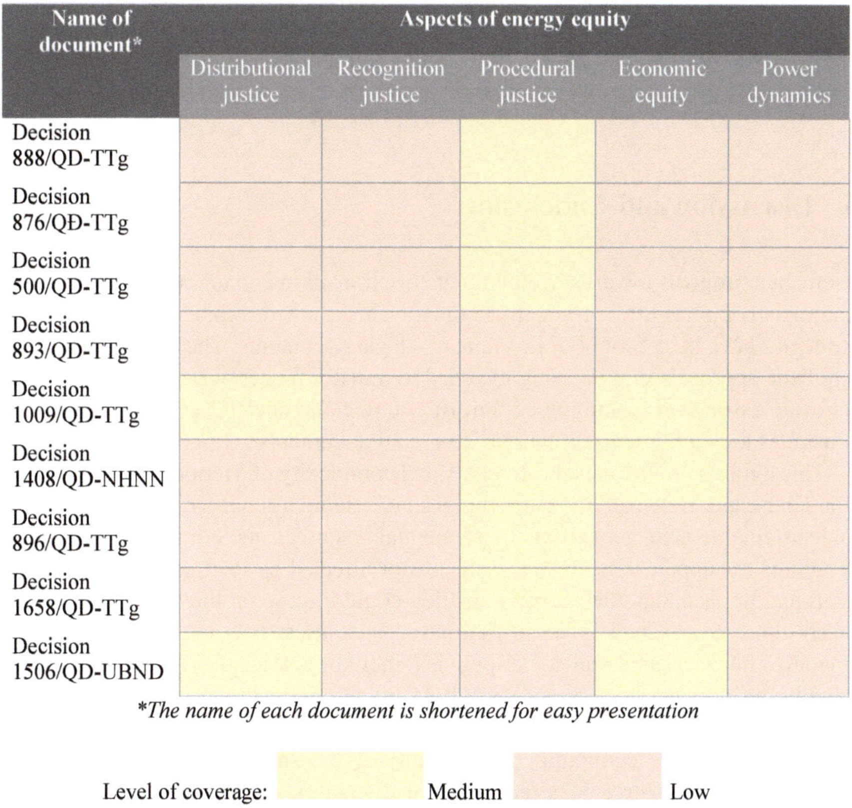

**Fig. 4** Heat map showing the extent to which each document covers the equity aspects

However, the extent of coverage was inadequate. The documents did not discuss clear mechanisms or provide guidance on how different stakeholders, including communities, could effectively partake in the energy transition decision-making process. There was, however, a focus on obtaining international support and involving business communities.

All documents recognized economic equity and focused on the transition to a green trajectory. Key aspects addressed included workforce development and incentives for businesses and communities. However, the depth of analysis varied, and specific support measures for those affected by economic changes were not outlined. Additionally, the documents did not delve into strategies for revitalizing areas heavily reliant on fossil fuels.

### 4.3   Power Dynamics

All documents failed to explicitly outline the specific power structures that might influence the implementation of strategies. This refers to the distribution of decision-making power and impact among different stakeholders, including governmental entities, industry players, associations, and communities. Additionally, there were no proposals to ensure engagement by marginalized or underrepresented groups.

## 5   Discussion and Conclusion

Vietnam's progress towards achieving energy transition is marked by various challenges and opportunities. To facilitate the transition, a wide range of approaches and strategies have been proposed in Vietnam's legal documents. These strategies adopt a holistic approach, emphasizing the need to transform energy sources as well as to innovate associated technologies, infrastructures, and societal behaviors. Figure 5 illustrates the eight recurrent themes across all documents.

This figure demonstrates the diversity and complexity of Vietnam's energy transition landscape. However, the current plans lack elements concerning the restoration and sustainable management of environmental resources, including land, and social programs to support those losing their jobs or affected by the transition. Given the profound impacts that the energy transition could impose on the workforce, particularly those entrenched in fossil-fuel-based industries, it is necessary to explore revitalization measures and social policies, including strategies for unemployment benefits, retirement schemes, and re-skilling and retraining initiatives (Bao Phap Luat 2023).

Additionally, the economic ramifications of the shift away from fossil fuels transcend the workforce to affect regional dynamics. Considering the geographical concentration of fossil fuel industries and renewable energy projects (Nguyen 2023), there is a significant need for an extensive regional development strategy to

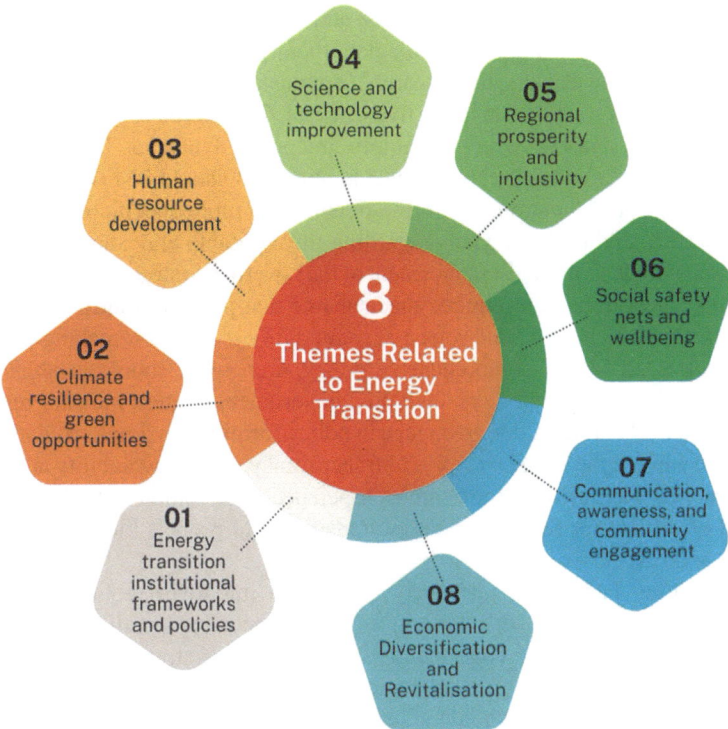

**Fig. 5** Eight energy transition themes across all studied documents

mitigate potential inequities, ensuring that all communities have access to economic opportunities arising from the transition. Economic diversification strategies must be developed to counteract the effects on regions relying on coal-related activities, directing them toward sustainable trajectories while preventing unwarranted economic damages.

Aligning with global commitments on net-zero emissions and achieving Vietnam's ambitious energy transition targets requires robust policy frameworks. Crucial considerations such as the intricacies of the shift away from fossil fuels, legal provisions for labor protection, and social security should be incorporated to ensure that affected communities are not disenfranchised and that their voices and rights are protected before, during, and after the transition (Carley and Konisky 2020). Another focus should be placed on integration and coordination across policy areas and policies (Wu et al. 2022). Energy transition strategies should be linked to other green growth and sustainability strategies and objectives to ensure effectiveness and alignment across the government's priorities, thus contributing to the avoidance of overlapping or contradictory policies. This also allows the government to systematically recognize intersections to ensure that policies are mutually reinforcing and effectively allocate resources. Facilitating the energy transition entails proactive and anticipatory measures for long-term gains, and it is pivotal that the government has adequate resources to initiate and implement them.

Regarding equity aspects, the research's appraisal was guided by the conceptual framework (Fig. 1), which includes five criteria: distributional justice, procedural justice, recognition justice, economic equity, and power dynamics. These emerged as crucial criteria necessitating careful consideration and should be prioritized. While these aspects found mention within the documents, more specific details, performance metrics, and concrete goals regarding the equitable allocation of resources, promotion of inclusivity, and enhancement of the governance structure were essential. Achieving a just transition requires mechanisms that promote transparency, inclusivity, citizen participation, and robust accountability in decision-making processes (Barlow et al. 2022). A comprehensive policy framework should establish clear conduits for diverse stakeholders, including government entities, industry representatives, and local communities, to engage in policy formulation, implementation, and evaluation while upholding principles of transparency and accountability at all levels of governance. The convergence of multiple perspectives promotes a holistic comprehension of difficulties related to energy transition and propels the development of relevant and responsive measures.

Furthermore, facilitating just energy transition initiatives requires multi-level government collaboration with a strong focus on supporting the affected local labor market and investing in the communities, including innovating and developing the local infrastructure in the affected regions. Emphasis should also be placed on collaboration with key stakeholders such as enterprise communities, particularly state-owned enterprises, and international partners. This collaboration can support Vietnam in securing the necessary funds for various aspects of the energy transition, including investments in business and sustainable energy programs, early retirement for coal plants, and retraining workers for green jobs.

Energy transition, as demonstrated through the conceptual framework (Fig. 2), is a co-evolutionary and dynamic process that unfolds in a non-linear manner and involves interconnected developments. In this context, Vietnam's journey toward achieving energy transition necessitates a meticulous balancing act, incorporating socio-economic-political conditions, regional discrepancies, and pertinent policy frameworks. It is pivotal that the energy transition discourse incorporates equity considerations and the development of policies extend beyond the traditional dichotomy of economic growth and environmental protection, laying a foundation for a just and equitable energy future.

# References

Ahlborg H, Nightingale AJ (2018) Theorizing power in political ecology: the 'where' of power in resource governance projects. J Polit Ecol 25(1):381–401. https://doi.org/10.2458/v25i1.22804

Bao Phap Luat (2023) Energy transition: coal labor and thermal power face numerous unforeseen risks. https://baophapluat.vn/chuyen-dich-nang-luong-lao-dong-than-nhiet-dien-than-doi-mat-nhieu-moi-nguy-co-khon-luong-post496702.html

Barlow J, Tapio R, Tarekegne B (2022) Advancing the state of energy equity metrics. Electr J 35(10):107208. https://doi.org/10.1016/j.tej.2022.107208

Bixler RP, Dell'Angelo J, Mfune O, Roba H (2015) The political ecology of participatory conservation: institutions and discourse. J Polit Ecol 22(1):164–182. https://doi.org/10.2458/v22i1.21083

Carley S, Konisky DM (2020) The justice and equity implications of the clean energy transition. Nat Energy 5:569–577. https://doi.org/10.1038/s41560-020-0641-6

Chen C, Feng J, Luke N, Kuo C-P, Fu JS (2022) Localized energy burden, concentrated disadvantage, and the feminization of energy poverty. Iscience 25(4):104139. https://doi.org/10.1016/j.isci.2022.104139

Cong PT, Manh DH, Huy HA, Phuong TT, Tuyen LT (2016) Livelihood vulnerability assessment to climate change at community level using household survey: a case study from Nam Dinh Province, Vietnam. Mediter J Soc Sci. https://doi.org/10.5901/mjss.2016.v7n3s1p358

Cook KS, Hegtvedt KA (1983) Distributive justice, equity, and equality. Ann Rev Sociol 9(1):217–241. https://doi.org/10.1146/annurev.so.09.080183.001245

Ember. Viet Nam. Attaining net zero for Viet Nam requires a ramp up of non-hydro power

Fitch Ratings (2022) Vietnam electricity. https://www.fitchratings.com/research/corporate-finance/vietnam-electricity-10-10-2022

Honneth A (2004) Recognition and justice. Acta Sociol 47(4):351–364. https://doi.org/10.1177/0001699304048668

Jenkins K, McCauley D, Heffron R, Stephan H, Rehner R (2016) Energy justice: a conceptual review. Energy Res Soc Sci 11:174–182. https://doi.org/10.1016/j.erss.2015.10.004

Lamont J (ed) (2017) Distributive justice. Routledge. https://doi.org/10.4324/9781315257563

Newell P (2008) The political economy of global environmental governance. Rev Int Stud 34(3):507–529. https://doi.org/10.1017/S0260210508008140

Nguyen KD, Ancev T, Randall A (2019) Evidence of climatic change in Vietnam: some implications for agricultural production. J Environ Manage 231:524–545. https://doi.org/10.1016/j.jenvman.2018.10.011

Nguyen T (2023) Vietnam's challenge to wean off coal. Lowy Institute. https://www.lowyinstitute.org/the-interpreter/vietnam-s-challenge-wean-coal

Osborne T, Brock S, Chazdon R, Chomba S, Garen E, Gutierrez V et al (2021) The political ecology playbook for ecosystem restoration: principles for effective, equitable, and transformative landscapes. Glob Environ Chang 70:102320. https://doi.org/10.1016/j.gloenvcha.2021.102320

Robertson M (2015) Environmental governance: political ecology and the state. In: The Routledge handbook of political ecology. Routledge, pp 457–466

Robbins P (2019) Political ecology: a critical introduction. Wiley

Schweiger G (2019) Recognition, misrecognition and justice. Ethics Glob Polit 12(4):1693870. https://doi.org/10.1080/16544951.2019.1693870

Schlosberg D, Collins LB (2014) From environmental to climate justice: climate change and the discourse of environmental justice. Wiley Interdiscipl Rev Clim Change 5(3):359–374. https://doi.org/10.1002/wcc.275

Sovacool BK (2021) Who are the victims of low-carbon transitions? Towards a political ecology of climate change mitigation. Energy Res Soc Sci 73:101916. https://doi.org/10.1016/j.erss.2021.101916

Sovacool BK, Dworkin MH (2015) Energy justice: conceptual insights and practical applications. Appl Energy 142:435–444. https://doi.org/10.1016/j.apenergy.2015.01.002

Stephens G (1986) Participatory justice: the politics of the future. Justice Q 3(1):67–82. https://doi.org/10.1080/07418828600088801

Swyngedouw E (2009) The political economy and political ecology of the hydro-social cycle. J Contemp Water Res Educ 142(1):56–60. https://doi.org/10.1111/j.1936-704X.2009.00054.x

Vietnamplus (2023) Vietnam stepping up int'l cooperation in climate change response. https://en.vietnamplus.vn/vietnam-stepping-up-intl-cooperation-in-climate-change-response/267438.vnp#:~:text=Notably%2C%20at%20the%2026th%20UN,to%20Clean%20Power%20Transition%20Statement

World Bank (2023) The World Bank in Vietnam. https://www.worldbank.org/en/country/vietnam/overview

World Bank Group and the Asian Development Bank (2021) Climate risk country profile: Vietnam

Wu J, Zuidema C, de Roo G (2022) Climate policy integration on energy transition: an analysis on Chinese cases at the local scale. Cities 120:103469. https://doi.org/10.1016/j.cities.2021.103469

# Motivators and Challenges of Ensuring Social Sustainability in the Apparel Industry in Bangladesh During COVID-19

**Shobod Deba Nath, Mohammad Mobarak Hossain, Farjana Nasrin, and Koushik Prashad Pathak**

**Abstract** This paper investigates the motivators and challenges in ensuring social sustainability practices in the apparel industry of Bangladesh during the COVID-19 pandemic. Utilizing a qualitative research design, the study conducted 30 in-depth semi-structured interviews with owners, managers, and workers within the industry to understand their vulnerabilities during the pandemic. The thematic analysis reveals that the COVID-19 pandemic has negatively impacted the apparel industry, particularly concerning workers' health issues, financial hardship, and the inability to afford essentials such as food and future employment opportunities. This situation has intensified the conflict between workers' livelihoods and health safety, posing a serious concern for all stakeholders in the apparel industry. Nonetheless, key motivators for maintaining social sustainability during the COVID-19 crisis include alleviating employees' psychological dilemmas about working during this period, attracting foreign buyers, and promoting sustainable innovation and protection against competition. Consequently, this paper offers valuable insights for owners, managers, and policymakers in the apparel industry, aiding them in implementing innovative techniques to ensure social sustainability practices during crises.

**Keywords** Social sustainability · Apparel industry · COVID-19 · Bangladesh · Worker safety · Compliance challenges

S. D. Nath · M. M. Hossain
Department of International Business, University of Dhaka, Dhaka, Bangladesh
e-mail: shobod@du.ac.bd

M. M. Hossain
e-mail: mobarak.ib@du.ac.bd

F. Nasrin (✉)
Department of Business and Technology Management, Islamic University of Technology (IUT), Gazipur City, Bangladesh
e-mail: farjana@iut-dhaka.edu

K. P. Pathak
Department of Marketing, University of Dhaka, Dhaka, Bangladesh
e-mail: kpp@du.ac.bd

© The Author(s), under exclusive license to Springer Nature Singapore Pte Ltd. 2024     113
A. Sharifi et al. (eds.), *Navigating Peace and Sustainability in an Increasingly Complex World*, World Sustainability Series, https://doi.org/10.1007/978-981-97-8772-2_6

# 1 Introduction

Ensuring social sustainability has recently garnered significant attention from industry leaders, academics, and policymakers worldwide (Khokhar et al. 2020), highlighting the need to understand the actual perception of social sustainability practices. Social sustainability encompasses the betterment of workers regarding job-related issues (job security, wages, work environment, allowances, and pension benefits) and health-related issues (workplace safety and security, medical support, among others) within an organization (Köksal et al. 2018; Ullah et al. 2021).

Additionally, internal and external drivers influence positive outcomes in the social sustainability of human factors (Sudusinghe and Seuring 2020). Company profit, brand image, compliance with government regulations, and adherence to foreign buyers' standards significantly impact social sustainability practices in apparel companies, ensuring long-term business sustainability (Anisul Huq et al. 2014). Köksal et al. (2017) emphasized that companies' internal strategies and intentions to provide social benefits are crucial triggers for ensuring social sustainability. Moreover, during the COVID-19 lockdown, employment assurance coupled with safety measures bolstered workers' mental strength, thereby promoting social sustainability practices in Bangladesh's apparel sector.

Some scholars have asserted that, although prioritizing social sustainability in the apparel industry is essential (Ashby et al. 2012), the uncertainty brought by the COVID-19 situation complicates this endeavor, especially in developing countries where ensuring social justice is even more critical (Khokhar et al. 2020). Other studies (Akbar and Ahsan 2021; Su et al. 2022) have identified challenges faced by companies, including regulatory compliance, resource management, shifting consumer demands, and cost issues. Apparel companies must meet foreign requirements to maintain full business operations, but this can be challenging due to workers' difficulties in complying with the rules. Additionally, a lack of knowledge about social sustainability, among workers and management often results in less attention to this issue.

During the COVID-19 pandemic, the cancellation of most foreign orders and factory shutdowns due to the virus's rapid spread exacerbated these challenges. Companies struggled to pay workers' full wages due to reduced profits and additional costs incurred for providing safety equipment. Thus, ensuring social sustainability in Bangladesh's apparel industry has become increasingly difficult during the pandemic.

This study addresses the following research questions: (1) What challenges did Bangladesh face regarding ensuring social sustainability in the apparel sector during the COVID-19 pandemic? (2) What factors motivate owners and managers to integrate social sustainability in the apparel industry during the COVID-19 pandemic?

To answer these questions, it is necessary to develop an in-depth understanding of social sustainability practices in the apparel industry and their real-world implications. The apparel sector is crucial for Bangladesh's economy, contributing more than 80% of the country's foreign trade earnings and supporting its competitive sustainability (Alam et al. 2019; Islam 2021; Rahman and Chowdhury 2020).

The study employs qualitative research, using convenience sampling techniques and semi-structured interviews to understand perceptions and realities of social sustainability practices. Data were collected from top-level, mid-level, and workers within the industry and analyzed using thematic analysis to uncover the real challenges and motivators during the COVID-19 crisis.

This paper makes several key contributions. First, it provides new insights for policymakers to implement sustainable social practices during crises such as the COVID-19 pandemic. Second, it identifies challenges that can help overcome post-pandemic issues in the apparel sector to ensure sustainable business practices. Finally, it highlights the real motivating factors and techniques for guaranteeing social sustainability during the pandemic.

The remainder of the paper is organized as follows. The literature review section focuses on relevant previous studies to gain prior insights into social sustainability issues, such as the conceptualization of social sustainability, challenges, and motivating factors, particularly in developing countries like Bangladesh. Next, the research methodology is described in Sect. 3. Sections 4 and 5 present the results and theoretical discussion of the findings. Finally, the last section outlines theoretical and practical implications and suggests future research areas.

## 2 Literature Review

### 2.1 Conceptualizing Social Sustainability

Sustainability refers to an organization's ability to ensure long-term betterment for society by enhancing economic value, involving all stakeholders, and safeguarding the environment (Eizenberg and Jabareen 2017; Tommasetti et al. 2020). Therefore, a primary dimension of sustainability is ensuring social sustainability from both community and industry development perspectives. More specifically, social sustainability entails sustainable practices in all aspects of human life, which ultimately lead to long-term human sustainability (Golicic et al. 2020).

The concept of social sustainability is largely based on closely related societal aspects, making its conceptualization complex and diversified (Weingaertner and Moberg 2014). Additionally, Lehtonen (2004) noted that this complexity creates vague areas in the practical implementation of social sustainability's theoretical concepts. Among the three primary sustainability indicators, economic and environmental sustainability have achieved broader acceptance and implementation, while social sustainability has yet to gain comparable momentum (Köksal et al. 2018; Mani et al. 2015). In non-Western countries, the debate over social sustainability remains indecisive due to a lack of comprehensive understanding of how to implement sustainability practices in all areas (Liu et al. 2017).

Social sustainability concerns the well-being of society and its people, which is a challenging task, especially during the COVID-19 pandemic in developing countries like Bangladesh (Awan et al. 2020). According to Khokhar et al. (2020) and Shen et al. (2011), social sustainability is multi-dimensional, encompassing urban sustainability, suppliers' and manufacturers' social sustainability, and customers' and companies' social sustainability. Additionally, Liu et al. (2017) discussed social sustainability from the perspective of social justice, including equality in rights, opportunities, and resource allocation, as well as overall well-being in human life. Thus, the concept of social sustainability is broadening to include the practices of communities and concerned institutions to fulfill human needs, dignity, and social justice (Jeekel 2017; Mesquita and Missimer 2021).

## 2.2 Challenges of Implementing Social Sustainability Practices in the Apparel Industry

Adopting social sustainability practices is challenging for all developing countries due to weaker institutional policies and a lack of understanding of the importance of sustainable development (Mani et al. 2018). Lehtonen (2004) found that the challenges of implementing sustainable practices are becoming more acute due to debates about trade-offs and synergies among different dimensions of social sustainability. In developing countries like Bangladesh, this task is particularly difficult as multi-tier suppliers and a large workforce are beset by numerous social issues (Mathiyazhagan et al. 2023). Additionally, implementing social sustainability in the apparel industry is complicated due to challenges in resource management, financial constraints, institutional weaknesses, regulatory hurdles, and inadequate monitoring systems (Akbar and Ahsan 2021; Awan et al. 2020).

Bangladesh faces the risk of losing a skilled labor force if apparel companies fail to maintain required social standards. Misalignments in the code of conduct can also lead to a reduction in foreign orders (Huq and Stevenson 2020). Grace Annapoorani (2017) highlighted several challenges in the Bangladeshi apparel industry, including workers' health and safety issues, abuse, gender inequality, wage discrimination, and child labor, which impede the proper implementation of social sustainability. Ensuring social sustainability practices among all multi-tier suppliers is particularly crucial for developing countries (Govindan et al. 2021).

## 2.3 Motivating Factors of Implementing Social Sustainability in the Apparel Industry

The factors driving firms to be socially sustainable are the motivators behind implementing social sustainability practices in the apparel industry. In today's business environment, it is common for all industries to strive for sustainable operations. As integral parts of society, firms are socially obligated to contribute to societal and individual betterment (Besser 2012; Falck and Heblich 2007). Companies endeavor to enhance workers' benefits by ensuring job security, health safety, and a favorable work environment to implement social sustainability in the apparel industry (Anisul Huq et al. 2014). Additionally, companies' internal strategies focused on social welfare issues are significant triggers for ensuring social sustainability in this sector (Köksal et al. 2017).

In the context of Bangladesh, a developing country, exporting quality textile products is a major source of foreign income and employment opportunities. Compliance with the exporting countries' code of conduct is crucial for building sustainable trust between buyers and suppliers, thereby enhancing social sustainability (Huq and Stevenson 2020). Oelze et al. (2020) noted that apparel companies are striving to ensure social sustainability by obtaining certification schemes, which also contribute to long-term business sustainability.

# 3   Methodology

## 3.1   Research Design

This study employs exploratory and qualitative research to investigate the motivators and challenges in ensuring social sustainability practices in the apparel industry of Bangladesh during the COVID-19 pandemic, a topic that remains insufficiently explored. Qualitative research aims to understand how people interpret social reality and the practical implementation of sustainability. Thus, this study will provide insights into the social sustainability issues, challenges, and motivating factors within Bangladesh's apparel industry (Nath and Eweje 2021; Voronov and Weber 2020). Additionally, thematic analysis has been applied for data analysis, offering flexibility in interpreting complex social issues (Kiger and Varpio 2020).

## 3.2   Data Collection

To understand the real scenario of implementing sustainable social practices, 30 respondents, including 25 managers and 5 workers, were interviewed. This survey was conducted from September 2021 to January 2022 using a convenience sampling technique, which provided the researcher with easier accessibility to data, especially during the COVID-19 crisis (Robinson 2014). The Dhaka district was chosen for data collection, focusing on key areas such as Gulshan, Zirabo, Keraniganj, Ashulia, Tejgaon, Narayanganj, and Gazipur, which are significant hubs in the apparel sector.

Identifying the most appropriate respondents was a prime concern in achieving the research objectives related to social sustainability in the apparel industry. Participants from top-level and mid-level management, as well as workers, were interviewed due to their influence on the company's internal policies and their broader impact on society.

During data collection, 30 in-depth interviews were conducted, each lasting between 40 and 90 min, with directors, supply chain managers, and workers from various export-oriented apparel companies. These interviews aimed to understand the practical scenario of social sustainability practices. Open discussions were facilitated to capture the respondents' perceptions and experiences regarding the challenges and motivating factors involved in integrating social sustainability into the apparel industry in Bangladesh (Table 1).

## 3.3  Data Analysis

To analyze the data set, thematic analysis has been employed to provide greater flexibility in interpreting textual data within a step-by-step qualitative research process. This approach allows for a deeper understanding of ideas and assumptions (Braun and Clarke 2006). To ensure data clarity, all interviews were transcribed line by line. A flexible data coding process was then used to identify perceptions about social sustainability, motivations behind sustainability practices, and the associated challenges. Semi-structured interviews guided the data collection and analysis process step by step.

To ensure the credibility and consistency of the thematic findings, supporting sample quotes from respondents were carefully explained throughout the data analysis process (Daniel 2018; Perry et al. 2015). This approach helped to validate the findings and provide a nuanced understanding of the participants' perspectives on social sustainability in the apparel industry of Bangladesh during the COVID-19 pandemic.

## 4  Results and Discussions

### 4.1  Challenges of Ensuring Social Sustainability in the Apparel Industry of Bangladesh

The COVID-19 pandemic, a significant threat to human existence, has also severely impacted economies worldwide. Bangladesh's economy, heavily reliant on export income and foreign direct investment driven by the apparel industry, is no exception (Islam 2021). The garment industry in Bangladesh accounts for more than 84% of total exports and employs over 4 million people across approximately 4,500 factories.

**Table 1** Respondent's profile

| SL | Designation | Company | SL | Designation | Company |
|----|-------------|---------|----|-------------|---------|
| 1 | Director (Supply chain) | Unisense Apparels LTD | 2 | Manager | One Tex BD Limited |
| 3 | Director (HR and CSR) | Unisense Apparels LTD | 4 | Deputy Manager | PK Knit Fashion |
| 5 | Senior Supply Chain Manager | Bextex Ltd. | 6 | HR Manager | Epyllion Group |
| 7 | Manager | Bextex Ltd. | 8 | Supply Chain Manager | Epyllion Group |
| 9 | HR Manager | Meghna Knit Composite Ltd. | 10 | Merchandiser | VIYELLATEX |
| 11 | Deputy Manager | Meghna Knit Composite Ltd. | 12 | HR Manager | VIYELLATEX |
| 13 | Supply Chain Manager | Unifill group | 14 | HR Manager | PK Knit Fashion |
| 15 | HR Manager | Unifill group | 16 | Manager | One Tex BD Limited |
| 17 | Merchandiser | Amtranet Group | 18 | Director (Supply chain) | Shasha Denims Limited |
| 19 | Director (HR and CSR) | Amtranet Group | 20 | Director (HR and CSR) | Shasha Denims Limited |
| 21 | Senior Supply Chain Manager | Grameen Knitwear Ltd. | 22 | Worker 1 | Unitex knitwear Ltd. |
| 23 | HR Manager | Grameen Knitwear Ltd. | 24 | Worker 2 | Bextex Ltd |
| 25 | HR Manager | Unitex knitwear Ltd. | 26 | Worker 3 | Unisense Apparels LTD |
| 27 | Director (Supply chain) | Knit Concern Group | 28 | Worker 4 | VIYELLATEX |
| 29 | Supply Chain Manager | ZEX Fashion Bangladesh | 30 | Worker 5 | PK Knit Fashion |

Unfortunately, this vital revenue-generating sector suffers from dire health and safety conditions for its employees, conditions further exacerbated by the pandemic.

The COVID-19 crisis has had a profoundly negative impact on garment workers, particularly in terms of health issues, financial hardship, inability to afford essentials such as food, and future employment opportunities. Ensuring stress-free work environments and implementing social sustainability practices is a daunting task for developing countries like Bangladesh, made even more complex by the pandemic. The major challenges in implementing social sustainability include the difficulty in meeting government and foreign buyers' requirements, ensuring environmental safety, lack of knowledge, factory shutdowns, workers' salaries, order cancellations, and the increased cost of providing safety equipment during the pandemic.

Figure 1 summarizes the conceptual framework and the thematic indicators, along with influencing factors based on the responses from the interviewees.

### *Incompatibility of Maintaining Government and Foreign Buyers' Requirements*

Apparel companies in Bangladesh often face challenges in maintaining government regulations concerning workers' safety, security, work environment, and training. These challenges are further compounded by the necessity to adhere to Western codes of conduct, which can be difficult due to cultural differences and trust issues (Anisul Huq et al. 2014). One interviewee expressed the situation as follows:

> To maintain foreign business relationships, companies must comply with numerous foreign buyers' requirements, especially during the COVID-19 pandemic, regarding employees' safety, security, health, training, work environment, and social sustainability. We focus on various issues like health, hygiene, dining space, medical facilities, doctor arrangements, a daycare center for workers' children, and different spaces in the office premises. However, we faced numerous challenges in ensuring these during the COVID-19 crisis. As a result, foreign buyers often did not respond positively.

Furthermore, challenges in the Bangladeshi apparel industry include ensuring quality assurance and cost efficiency aligned with labor standards, which were already problematic due to operational inefficiencies before the pandemic. The onset of the COVID-19 crisis made compliance with health and safety regulations, PPE

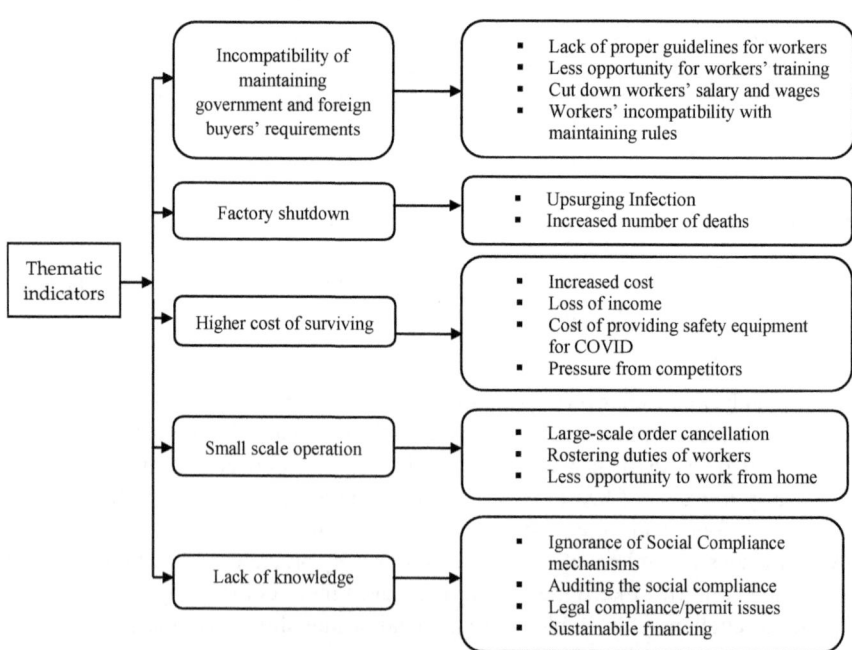

**Fig. 1** Conceptual framework and indicators of the Challenges of ensuring social sustainability in the apparel industry of Bangladesh, especially in a covid pandemic situation

provisions, and social distancing measures more critical, exacerbating existing challenges (Majumdar et al. 2020). The lack of control over global demand fluctuations, supply chain disruptions, and the incompatibility of maintaining both government and foreign buyers' requirements have acted as significant barriers to ensuring social sustainability practices in the apparel industry during the COVID-19 crisis.

The following sections expound on several influencing factors of these challenges as perceived by the interviewees.

### 1. Lack of Proper Guidelines for Workers

Bangladesh, as a developing country, often faces difficulties in establishing proper guidelines in the apparel sector due to misalignments between foreign and local requirements (Anisul Huq et al. 2014). Additionally, the global apparel industry operates with multiple layers of suppliers and subcontractors, posing a significant challenge for Bangladesh in monitoring the entire supply chain. Respondents from the company level indicated that inconsistencies in the regulatory framework governing the apparel sector in Bangladesh further complicate this issue.

The misalignments between national and international guidelines create an even more complex scenario during crises like the COVID-19 pandemic, making it difficult for apparel companies to ensure social sustainability practices (Hasan et al. 2021). Managers and supervisors faced significant challenges in supervising workers, leading to serious mismanagement in several apparel companies. One supply chain manager stated:

> Most of the apparel companies face a lack of proper guidelines and international standards for workers' safety, security, wages, and other precautionary measures. Hence, understanding social sustainability issues is far beyond the thinking of workers and managers, who do not receive proper regulatory and monitoring guidelines.

### 2. Less Opportunity for Workers' Training

Workers in the apparel industry in Bangladesh usually do not receive proper training facilities. From the company's perspective, managers are reluctant to incur the costs associated with providing training facilities, as they find it less impactful for the company in the context of global competition. Similar findings were reported by Akbar and Ahsan (2021), who emphasized cost issues, resource utilization, and financial concerns as factors resulting in fewer training arrangements for workers. Several workers stated:

> We don't get the proper training and development to improve our working skills and productivity. Sometimes we fear accidents and feel insecure because we haven't received any safety and rescue training.

Moreover, workers lack training on safety concerns (James et al. 2019). These issues make social sustainability more challenging for managers and workers in apparel companies, a problem that was exacerbated during the COVID-19 pandemic.

### 3. *Reduction of Workers' Salaries and Wages*

During the COVID-19 pandemic, apparel companies faced high production costs and large order cancellations, leading to cuts in workers' salaries to reduce operational costs. Apparel companies worldwide faced similar challenges, with Bangladesh experiencing severe impacts due to its status as a developing country (Helppie-McFall and Hsu 2020). Additionally, Syed (2024) found that COVID-19 created significant challenges in ensuring social sustainability due to the difficulty of maintaining policies and regulations. Companies were also unable to ensure the minimum wage for workers as stipulated in the revised amendment of the Bangladesh Labor Law in 2015. From the employer's perspective:

> Most of the apparel companies were facing losses due to COVID-19 shutdowns. Reducing the company's losses was our primary concern during the crisis, and we were bound to cut workers' salaries.

### 4. *Workers' Inability to Maintain Compliance*

Most apparel workers in Bangladesh face significant disparities in socioeconomic conditions, leading them to view rule compliance as optional rather than obligatory. Additionally, a lack of employment opportunities and training, inadequate enforcement mechanisms, and weak regulatory oversight and monitoring are the main reasons for workers' inability to maintain compliance (Narula 2019). Rahman and Rahman (2020) found that after the Rana Plaza and Tazreen Fashion incidents, it became crucial for all stakeholders to adhere to foreign regulations for international business. Several managers shared their experiences, noting:

> Usually, workers are not concerned about maintaining rules, even though they have been guided and equipped with all necessary items for their safety. We regularly inspect and monitor the working conditions, especially during the COVID-19 situation. However, without the workers' cooperation, it is hard to maintain rules and regulations in every aspect of ensuring social sustainability in Bangladesh.

There is a significant lack of trust between companies and workers during crises, creating a complex situation during the COVID-19 pandemic. This mistrust further complicates efforts to ensure social sustainability and effective rule compliance.

### *Factory Shutdowns*

The rate of COVID-19 infection and the number of deaths were rising uncontrollably, leading to the full or partial shutdown of factories. Consequently, most foreign orders were canceled. Apparel companies faced drastic declines in demand and higher production costs due to the surge in infections worldwide (Sim 2020). Running the business during these crisis moments became a major challenge, and ensuring social sustainability was beyond consideration. All managers and directors expressed similar sentiments during the interviews. One manager stated:

> During the period of COVID-19, we had to shut down our factories and halt our business operations to stop the spread of the virus. We value workers' lives, which are more important than any other aspect of a company's survival.

Nearly all companies faced rising cases of COVID-19 infections. To control the direct impact of sickness, factories had to shut down fully or partially based on the severity of the COVID-19 situation (Ashby et al. 2012; Castañeda-Navarrete et al. 2021; Rahman 2022; Rimi 2022).

### *Higher Survival Costs*

1. *Increased Survival Costs and Loss of Income for Apparel Companies*

The apparel industry in Bangladesh faced significant challenges due to higher survival costs during the COVID-19 pandemic, largely caused by the disruption of the global supply chain. Additionally, the shortage of raw materials further increased production costs (Sen et al. 2020). This disruption led to higher production costs during and after the pandemic, as the demand and consumption of non-essential goods decreased. Consequently, Bangladesh experienced severe income losses (Mia et al. 2020). Providing salaries to all employees became a major challenge due to reduced profits and deferred payments from foreign buyers and retailers (Rahman and Chowdhury 2020). As a result, many apparel companies reduced workers' wages by a percentage (Hasan et al. 2021). During the pandemic, companies incurred additional costs by providing masks, hand sanitizers, PPE, and transportation facilities. Furthermore, competitive challenges weakened the market position of apparel companies both locally and globally, as reported by several managers interviewed. One manager shared:

> During the COVID-19 situation, our normal business procedures were disrupted, leading to decreased income and profit. We had to pay workers' wages as a percentage. We were also concerned about ensuring the livelihood of all employees, providing masks, hand sanitizers, PPE, transportation facilities, and many other amenities. We faced the severe challenge of increased costs, and at that time, we had to focus more on our sustainability implementation practices.

This scenario highlights the compounded difficulties faced by the Bangladeshi apparel industry during the pandemic, where maintaining business operations and ensuring social sustainability practices became exceedingly challenging.

2. *Competitive Pressure*

As the second-largest clothing exporter, Bangladesh's apparel industry faces high risks in maintaining its significance in the global market due to competition from countries such as China, Vietnam, and Cambodia (Huq and Stevenson 2020). Anner (2020) found that the purchasing practices of raw materials by other global lead firms adversely affect developing countries by creating volatile orders, short lead times, and excessive pricing pressures. Several managers from the apparel companies described their situation during interviews. One manager stated:

> We are facing an invisible challenge of maintaining our presence in the global marketplace and fearing the loss of foreign orders due to the aggressive strategies of our competitors operating in China and Vietnam. During this crucial moment of the COVID-19 pandemic, if we are unable to provide better purchasing options to our foreign buyers, Bangladesh's apparel sector may suffer significant losses in the near future.

Therefore, Bangladesh's apparel industry faces severe challenges from competitors in other countries in terms of ensuring quality, maintaining an efficient supply chain, and sustaining operations while navigating unpredictable situations.

### Small-Scale Operations of Apparel Companies

The global apparel supply chain faced the most challenging business situation during the COVID-19 pandemic, compounded by different infectious COVID variants. Due to large-scale order cancellations from foreign buyers, apparel companies had to partially or fully shut down their operations in Bangladesh. Hasan et al. (2021) and McMaster et al. (2020) opined that decreased consumption patterns reduced the demand for apparel goods worldwide. Additionally, there were fewer opportunities to work from home in the apparel sector, which required companies to roster workers' duties. This strategy helped companies mitigate cost-cutting challenges. As a manager noted:

> Our maximum orders were halted but not canceled. We resumed operations on a small scale and implemented a roster-based work schedule. At the management level, work from home was implemented and made mandatory.

Since the impacts of COVID-19 in the apparel industry are interrelated and disrupt the global apparel supply chain, Bangladesh faces significant challenges in ensuring social sustainability in all aspects of the apparel industry.

### Lack of Knowledge

Workers' ignorance of legal and social compliance is common in Bangladesh due to a lack of knowledge about the importance of compliance mechanisms (Terwindt and Armstrong 2019). Moreover, the social and economic conditions of Bangladeshi workers in the apparel sector are very underprivileged compared to employees in other industries. As a result, workers are often unaware of their rights and health and safety regulations, which are essential for implementing social sustainability practices. This ignorance has created a culture of non-compliance, undermining efforts to implement social sustainability practices in Bangladesh (Reza and Du Plessis 2022).

To attract and retain foreign buyers, permit issues became more important during the COVID-19 situation as more inspections were conducted (Islam and Halim 2022). This increased scrutiny created a reluctance among companies to undergo auditing and ensure compliance with legal and social mechanisms. Consequently, it has become a paradox and a significant challenge for apparel companies to implement social sustainability practices, especially since foreign buyers are now more concerned about sustainability issues in the post-COVID-19 era. The lack of knowledge among workers about legal compliance, safety issues, sustainable financing, and other compliance mechanisms further complicates this challenge (Mia et al. 2020). This was supported by many interviewees who indicated that apparel companies sometimes face challenges due to mock compliance, as identified by Anisul Huq et al. (2014).

Before COVID-19, a lack of knowledge and awareness regarding international labor standards, environmental regulations, management best practices, and social issues among both employers and workers created significant challenges in ensuring social sustainability (Syed 2024). During the pandemic, employers faced additional complexities in implementing strict health and safety protocols, while workers struggled to adhere to these protocols, perceiving them as less important than ensuring their livelihoods. This knowledge gap between employers and workers makes the implementation of social sustainability practices a challenging issue for the apparel industry in Bangladesh. Several managers of apparel companies stated: "The biggest challenge is the lack of knowledge among workers about social sustainability, which is responsible for their ignorance of social compliance."

Additionally, the Bangladesh apparel sector faces inefficiencies in sustainable financing options due to the lack of knowledge and competencies among employers and company owners (Akbar and Ahsan 2021). Typically, company owners focus on short-term profits without considering long-term sustainability and efficiency in the apparel value chain. As a result, it has become difficult for apparel companies to cope with the COVID-19 crisis and implement social sustainability in the long run.

The disruption caused by the pandemic led to reduced production, and companies had to survive with minimal sales. Furthermore, due to a lack of coordination between the apparel sector and the government, factories did not reopen promptly, causing immense suffering for garment workers. Companies also incurred additional costs by providing safety equipment and ensuring social distancing in the workplace, particularly in congested areas. Workers faced a dilemma between working with health risks and staying home without a job. Based on workers' responses, they felt a significant life risk of COVID-19 infection due to having fewer savings and poor economic conditions to afford treatment. They also worried about who would take responsibility for their family's income, livelihood expenses, and children. Workers experienced extra mental pressure from the fear of losing jobs and receiving lower wages if they did not go to work.

One of the major challenges for Bangladesh is preventing further transmission among vulnerable populations. Factory owners have often prioritized protecting their livelihoods over their workers' lives by reopening factories when the pandemic was still at its peak. This has affected top-level management decisions regarding employees and threatens the implementation of social sustainability practices in the apparel sector in Bangladesh. In some cases, the impact of these decisions lingers even after the pandemic.

### *4.2 Factors Motivating Owners and Managers to Integrate Social Sustainability Practices in the Apparel Industry in Bangladesh During the COVID-19 Pandemic*

The motivating factors for apparel companies in Bangladesh to integrate social sustainability practices during the COVID-19 pandemic are explained in this section, based on the analysis of interviewees' opinions presented in Fig. 2. Most respondents identified key motivators that drive top-level managers to adopt policies aimed at implementing social sustainability practices in their organizations.

Several motivating factors drive company leaders in the apparel industry to adopt social sustainability practices. The most influential motivator is the desire to contribute to the development of society and its people. This includes providing employment opportunities, ensuring safety and security, offering retirement benefits, arranging employee training programs, and providing mental health support during crisis moments. Additionally, motivating employees towards societal betterment is crucial. These findings are supported by the perceptions of interviewees and corroborated by studies from Agradi et al. (2022), Anisul Huq et al. (2014), and Stamm (2023).

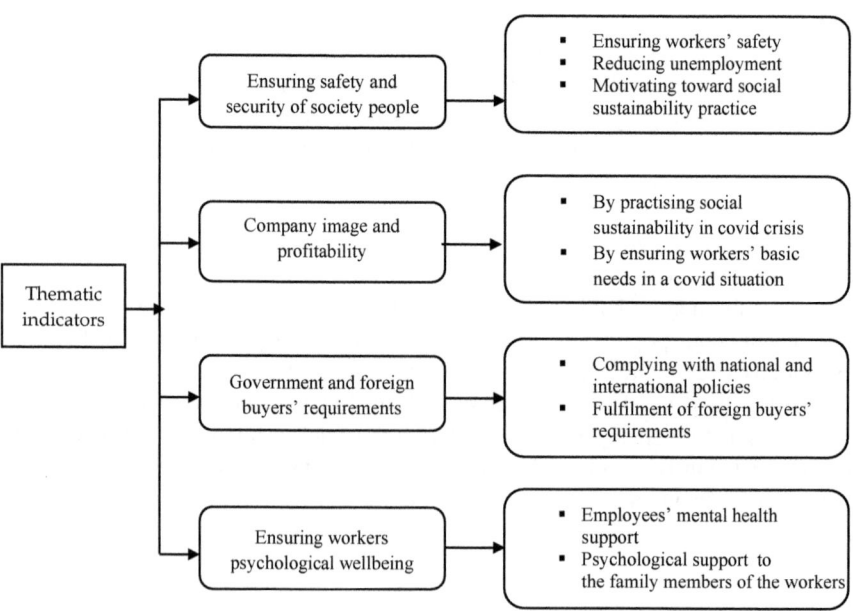

**Fig. 2** Motivating factors of implementing social sustainability practice in the apparel industry in Bangladesh

### Ensuring Worker Safety and Security

Apparel companies feel responsible for ensuring the safety and security of their workers, which is a primary motivator for adopting social sustainability practices. Nearly all apparel companies have tried to provide safety equipment, hand sanitizers, gloves, masks, and PPE materials to reduce the rate of infection (Gereffi 2020). At the same time, many companies have committed to securing workers' jobs despite the significant cost pressures during the crisis. Several respondents expressed similar thoughts:

> As part of society, the responsibility behind social development motivates us to contribute to societal well-being, especially in the apparel industry where many poor people are employed. Ensuring workers' safety, providing masks, hand sanitizers, soap, and PPE during COVID-19 is crucial. We must focus on employment opportunities and employee training because workers should be cautious about social sustainability issues.

Reducing unemployment during the COVID-19 crisis acts as another motivating factor since a large number of workers are losing their jobs. Ensuring job security for low-wage workers in crisis moments is a challenge, but apparel companies in Bangladesh take this challenge as a motivating factor to ensure social sustainability practices. They also emphasize retaining better employees by providing social security and standards. One respondent said:

> Supporting employees by assuring their job security and pay benefits during the COVID-19 crisis also acts as a motivator for the company to retain core personnel. Without the workers' progress, even if we achieve success, it will not be very meaningful to us. We believe this type of progress is not sustainable unless we focus on sustainable business practices.

Additionally, motivating employees towards social sustainability practices also leverages the benefits of secured jobs, especially since foreign buyers are highly concerned about whether social sustainability practices are implemented in developing countries, particularly after the Rana Plaza collapse in 2013 (Huq and Stevenson 2020; Siddiqui and Uddin 2016). Consequently, maintaining sustainable practices and providing a safe working environment has become more challenging, which ultimately acts as an important motivator for apparel companies in Bangladesh.

### Company Image and Profitability

Since the apparel industry of Bangladesh significantly impacts the country's economy and growth, its image in the global marketplace is also under constant scrutiny. By fostering positive relationships with workers, supply chain intermediaries, and foreign buyers, a company can maintain its image and profitability, which ultimately motivates apparel companies to adopt socially sustainable practices (Chiesa and Przychodzen 2019).

Based on respondents' perceptions, a company is considered to be doing well in social sustainability if it can ensure workers' safety and security measures, even in crisis moments. This perception can motivate company owners and managers to work for the betterment of society, which ultimately enhances the company's economic benefits and reputation. Additionally, practicing social sustainability is a primary concern for companies to ensure compliance with international standards and regulations, which acts as a motivating factor for the apparel industry in Bangladesh. One respondent noted:

> In this pandemic situation, a large pool of apparel workers are losing their jobs or receiving reduced wages. The reason behind this is the lack of sufficient profit from business and the extra cost of running a factory during the pandemic. So, in this crisis, if a company can ensure facilities, safety, and security, its brand image and reputation will be higher in the market.

Thus, company reputation and brand image are major driving factors of social sustainability, as indicated by interviewee responses and aligned with findings from several studies (Cooper et al. 2016; Epstein et al. 2018; Hutchins and Sutherland 2008).

### Government and Foreign Buyers' Requirements

Another motivating factor for ensuring social sustainability in the apparel industry of Bangladesh is the need to comply with national and international policies and meet foreign buyers' requirements. The significance of maintaining international policies has gained much attention following the Tazreen Fashion and Rana Plaza incidents in 2012 and 2013. Azizul Islam and Deegan (2008) and Hoque and Shahinuzzaman (2021) have argued that adhering to institutional requirements drives the apparel sector to boost demand in the global market, a demand that was at risk during the COVID-19 pandemic. While these requirements pose challenges during crises due to uncertain demand from other countries, they also act as motivating factors for apparel companies to ensure social sustainability. Company owners and managers expressed:

> Since Bangladesh heavily relies on apparel exports, a significant portion of our garment products are designed based on foreign requirements, including quality, quantity, lead time, and social and legal compliance mechanisms. In this regard, before COVID, during COVID, and after COVID, apparel companies have to maintain government and foreign buyers' requirements for sustainable business.

From the government's viewpoint, there is pressure from importers to continue apparel production, and the government expects employees to work to fulfill buyers' requirements. Company owners perceive that failure to comply with national and international policies may result in legal issues or trade sanctions. If local apparel companies can adapt to changing preferences in critical situations, maintain production and delivery capabilities, and build long-term relationships with foreign buyers, their business opportunities will enhance in the post-pandemic landscape. Thus, by maintaining government and foreign buyers' requirements, an apparel company demonstrates its commitment to social sustainability, even in crisis moments, which acts as a motivating factor to integrate these practices during the COVID-19 crisis.

### Ensuring Psychological Well-Being of Workers and Their Families

Workers in the apparel industry faced a complex psychological dilemma during the COVID-19 crisis. On one hand, they risked losing their jobs; on the other, they faced significant health and life risks due to the pandemic. Additionally, a large number of apparel workers moved back to their village homes after losing their jobs (Khanna 2020). Workers' family members often felt it was safer to stay in the village rather than face high life risks with no job security in urban areas during the crisis (Sakamoto et al. 2020). This perception among workers has prompted managers and owners of apparel companies to implement social sustainability practices more rigorously. According to workers:

> We are workers with minimum income and a single source of livelihood. If we cannot go to work, our families will suffer. However, if we return to work, we may get infected, putting our lives at stake. Sometimes companies open without proper safety measures. If companies can ensure our health safety and cover medical expenses, it will relieve our mental stress and support our family livelihoods.

Workers have shared various issues and experiences from different perspectives. Although Bangladesh has been highlighted for garment worker safety due to numerous past disasters (Bair et al. 2020), workers often feel neglected, particularly regarding their salaries. Workers from different factories in Savar, Ashulia, and Mirpur reported inadequate health and safety measures and little change in their working conditions. Many garment factories do not maintain social distancing guidelines on the working floor and at factory entrances, increasing the risk of contracting the coronavirus. Reports indicate that some factories dismissed workers infected with COVID-19 and laid off others, citing canceled orders from international buyers, which was not always true.

Companies can mitigate these challenges by focusing on preparedness, such as ensuring available treatment facilities, human capacity, and an emergency response system in Bangladesh. Publicly committing to these measures would significantly motivate workers to return to their jobs with less mental stress. Given the challenges faced by the global apparel supply chain during the COVID-19 crisis, apparel companies in Bangladesh view these issues as motivating factors to ensure social sustainability practices. Ensuring the psychological well-being of workers and their families can foster a sense of safety and loyalty among workers. Additionally, the well-being of workers' families positively influences the broader community and society, ultimately benefiting the business.

From the workers' viewpoint, companies should address employees' concerns and dire needs during the COVID-19 pandemic. Key points motivating apparel companies to integrate social sustainability during the crisis include ensuring workplace security and safety by providing masks, gloves, hand sanitizers, soap, PPE, entrance spray, temperature checking machines, and transportation facilities. Providing economic benefits, wages, Eid bonuses, and medical support will add value for the company, helping to retain customers and meet workers' expectations. These measures will motivate the integration of social sustainability in the apparel industry during the COVID-19 pandemic.

# 5  Conclusion

This paper investigated two research questions: (1) What challenges did Bangladesh face during the COVID-19 pandemic in the apparel sector regarding the implementation of social sustainability practices? (2) What factors motivate owners and managers to integrate social sustainability in the apparel industry during the COVID-19 pandemic? The findings revealed several serious problems, including order cancellations, factory shutdowns, increased costs for providing safety equipment, maintaining foreign requirements, and transportation disruptions. These challenges made it harder to achieve social sustainability practices during the COVID-19 situation in Bangladesh. By analyzing pre-COVID and COVID scenarios, this study aimed to minimize the gap between company owners and workers in understanding social sustainability practices.

In addition, this study highlights the interviewees' perceptions about economic and social development, supporting workers' mental health conditions, compliance with foreign buyers' requirements, and employee benefits, which act as motivating factors for social sustainability practices. Despite Bangladesh's highest efforts, operational inefficiencies and financial constraints in a developing country make it more complex to uphold the consistency of foreign policy compliance. As a result, the apparel sector in Bangladesh faces dual pressure: surviving in the world market while ensuring workers' welfare. Moreover, implementing social sustainability in the apparel sector not only increases societal benefits but also enhances the company's resilience and reputation in the global marketplace.

## 5.1  Theoretical Implications

This study makes significant contributions to the existing literature on social sustainable practices in the apparel industry. Although research on social sustainability in this sector is evolving, most studies focus on developed countries (Bubicz et al. 2021; Chiesa and Przychodzen 2019; Sudusinghe and Seuring 2020). This research highlights the challenges and motivators of ensuring social sustainability practices in Bangladesh's apparel industry, thereby enriching the existing knowledge in the field of social sustainability in developing countries. By integrating interviewees' perceptions and existing literature, this study explores the complexities arising from regulatory inconsistencies, inadequate worker training, and compliance issues. Furthermore, this research uncovers the gap between theoretical frameworks and the underlying realities faced by the apparel industry in developing countries like Bangladesh, where implementing social sustainability is particularly complex.

## 5.2 Implications for Managers and Policymakers

The findings of this study offer several valuable implications for policymakers, stakeholders, and employees. It reveals the challenges and motivations of social sustainability to company owners, managers, and decision-makers who aim to improve sustainability practices during crises. Understanding the major challenges and how these challenges can act as motivating factors—such as maintaining buyers' requirements, addressing legal permit issues, adhering to compliance mechanisms, and ensuring adequate knowledge—can help owners and managers of apparel companies implement social sustainability practices more effectively.

Additionally, ensuring workplace safety, supporting workers' psychological well-being, and practicing social sustainability can strengthen the competitive advantage of apparel companies in the global marketplace. Policymakers, including governments and trade associations, should be more aware of the potential threats and consequences of critical scenarios. By addressing these issues proactively, they can foster a more resilient and sustainable apparel industry in Bangladesh.

## 5.3 Future Research Directions

This research has some shortcomings in data collection, as most interviewees were from middle or top-level management. This limitation affects the generalizability of the findings. Future research could be extended to include other industries and other developing countries, along with a broader range of responses from the apparel sector. Additionally, empirically testing the findings with quantitative data could enrich and validate the results of this study.

**Informed Consent Statement**: Participants were well informed about the interview, and they agreed with us to share their practical experience in their working areas in the apparel companies of Bangladesh.

**Conflicts of Interests** :There is no conflict of interest.

# References

Agradi M, Adom PK, Vezzulli A (2022) Towards sustainability: does energy efficiency reduce unemployment in African societies? Sustain Cities Soc 79:103683. https://doi.org/10.1016/j.scs.2022.103683

Akbar S, Ahsan K (2021) Investigation of the challenges of implementing social sustainability initiatives: a case study of the apparel industry. Soc Respons J 17(3):343–366. https://doi.org/10.1108/SRJ-09-2019-0291

Alam MS, Selvanathan EA, Selvanathan S, Hossain M (2019) The apparel industry in the post-multifiber arrangement environment: a review. Rev Dev Econ 23(1):454–474. https://doi.org/10.1111/rode.12556

Anisul Huq F, Stevenson M, Zorzini M (2014) Social sustainability in developing country suppliers. Int J Oper Prod Manag 34(5):610–638. https://doi.org/10.1108/IJOPM-10-2012-0467

Anner M (2020) Squeezing workers' rights in global supply chains: purchasing practices in the Bangladesh garment export sector in comparative perspective. Rev Int Polit Econ 27(2):320–347. https://doi.org/10.1080/09692290.2019.1625426

Ashby A, Leat M, Hudson-Smith M (2012) Making connections: a review of supply chain management and sustainability literature. Suppl Chain Manage Int J 17(5):497–516. https://doi.org/10.1108/13598541211258573

Awan U, Kraslawski A, Huiskonen J, Suleman N (2020) Exploring the locus of social sustainability implementation: a south Asian perspective on planning for sustainable development, pp 89–105. https://doi.org/10.1007/978-3-030-30306-8_5

Azizul Islam M, Deegan C (2008) Motivations for an organisation within a developing country to report social responsibility information. Account Audit Account J 21(6):850–874. https://doi.org/10.1108/09513570810893272

Bair J, Anner M, Blasi J (2020) The political economy of private and public regulation in post-Rana Plaza Bangladesh. ILR Rev 73(4):969–994. https://doi.org/10.1177/0019793920925424

Besser TL (2012) The consequences of social responsibility for small business owners in small towns. Bus Ethic: Eur Rev 21(2):129–139. https://doi.org/10.1111/j.1467-8608.2011.01649.x

Braun V, Clarke V (2006) Using thematic analysis in psychology. Qual Res Psychol 3(2):77–101. https://doi.org/10.1191/1478088706qp063oa

Bubicz ME, Dias Barbosa-Póvoa APF, Carvalho A (2021) Social sustainability management in the apparel supply chains. J Clean Prod 280:124214. https://doi.org/10.1016/j.jclepro.2020.124214

Castañeda-Navarrete J, Hauge J, López-Gómez C (2021) COVID-19's impacts on global value chains, as seen in the apparel industry. Dev Policy Rev 39(6):953–970. https://doi.org/10.1111/dpr.12539

Chiesa PJ, Przychodzen W (2019) Social sustainability in supply chains: a review. Soc Respons J 16(8):1125–1148. https://doi.org/10.1108/SRJ-11-2018-0301

Cooper J, Stamford L, Azapagic A (2016) Shale gas: a review of the economic, environmental, and social sustainability. Energ Technol 4(7):772–792. https://doi.org/10.1002/ente.201500464

Daniel BK (2018) Empirical verification of the "TACT" framework for teaching rigour in qualitative research methodology. Qual Res J 18(3):262–275. https://doi.org/10.1108/QRJ-D-17-00012

Eizenberg E, Jabareen Y (2017) Social sustainability: a new conceptual framework. Sustainability 9(1):68. https://doi.org/10.3390/su9010068

Epstein MJ, Elkington J, Leonard HB, "Dutch" (2018) Making sustainability work. Routledge. https://doi.org/10.4324/9781351280129

Falck O, Heblich S (2007) Corporate social responsibility: doing well by doing good. Bus Horiz 50(3):247–254. https://doi.org/10.1016/j.bushor.2006.12.002

Gereffi G (2020) What does the COVID-19 pandemic teach us about global value chains? The case of medical supplies. J Int Bus Policy 3(3):287–301. https://doi.org/10.1057/s42214-020-00062-w

Golicic SL, Lenk MM, Hazen BT (2020) A global meaning of supply chain social sustainability. Prod Plan Control 31(11–12):988–1004. https://doi.org/10.1080/09537287.2019.1695911

Govindan K, Shaw M, Majumdar A (2021) Social sustainability tensions in multi-tier supply chain: a systematic literature review towards conceptual framework development. J Clean Prod 279:123075. https://doi.org/10.1016/j.jclepro.2020.123075

Grace Annapoorani S (2017) Social sustainability in textile industry, pp 57–78. https://doi.org/10.1007/978-981-10-2639-3_4

Hasan NA, Heal RD, Bashar A, Bablee AL, Haque MM (2021) Impacts of COVID-19 on the finfish aquaculture industry of Bangladesh: a case study. Mar Policy 130:104577. https://doi.org/10.1016/j.marpol.2021.104577

Helppie-McFall B, Hsu JW (2020) Financial profiles of workers most vulnerable to "coronavirus-related" earnings loss in the spring of 2020. Financ Plann Rev 3(4). https://doi.org/10.1002/cfp2.1102

Hoque I, Shahinuzzaman Md (2021) Task performance and occupational health and safety management systems in the garment industry of Bangladesh. Int J Workplace Health Manag 14(4):369–385. https://doi.org/10.1108/IJWHM-09-2020-0169

Huq FA, Stevenson M (2020) Implementing socially sustainable practices in challenging institutional contexts: building theory from seven developing country supplier cases. J Bus Ethics 161(2):415–442. https://doi.org/10.1007/s10551-018-3951-x

Hutchins MJ, Sutherland JW (2008) An exploration of measures of social sustainability and their application to supply chain decisions. J Clean Prod 16(15):1688–1698. https://doi.org/10.1016/j.jclepro.2008.06.001

Islam MdS (2021) Ready-made garments exports earning and its contribution to economic growth in Bangladesh. GeoJournal 86(3):1301–1309. https://doi.org/10.1007/s10708-019-10131-0

Islam T, Halim MdA (2022) Impact of ready-made garments (RMG) industries and sustainability: Perspective of the pandemic period in developing country. Clean Eng Technol 11:100567. https://doi.org/10.1016/j.clet.2022.100567

James P, Miles L, Croucher R, Houssart M (2019) Regulating factory safety in the Bangladeshi garment industry. Regul Govern 13(3):431–444. https://doi.org/10.1111/rego.12183

Jeekel H (2017) Social sustainability and smart mobility: exploring the relationship. Trans Res Procedia 25:4296–4310. https://doi.org/10.1016/j.trpro.2017.05.254

Khanna A (2020) Impact of migration of labour force due to global COVID-19 Pandemic with reference to India. J Health Manag 22(2):181–191. https://doi.org/10.1177/0972063420935542

Khokhar M, Iqbal W, Hou Y, Abbas M, Fatima A (2020) Assessing supply chain performance from the perspective of Pakistan's manufacturing industry through social sustainability. Processes 8(9):1064. https://doi.org/10.3390/pr8091064

Kiger ME, Varpio L (2020) Thematic analysis of qualitative data: AMEE Guide No. 131. Med Teacher 42(8):846–854. https://doi.org/10.1080/0142159X.2020.1755030

Köksal D, Strähle J, Müller M (2018) Social sustainability in apparel supply chains—the role of the sourcing intermediary in a developing country. Sustainability 10(4):1039. https://doi.org/10.3390/su10041039

Köksal D, Strähle J, Müller M, Freise M (2017) Social sustainable supply chain management in the textile and apparel industry—a literature review. Sustainability 9(1):100. https://doi.org/10.3390/su9010100

Lehtonen M (2004) The environmental–social interface of sustainable development: capabilities, social capital, institutions. Ecol Econ 49(2):199–214. https://doi.org/10.1016/j.ecolecon.2004.03.019

Liu Y, Dijst M, Geertman S, Cui C (2017) Social sustainability in an ageing Chinese society: towards an integrative conceptual framework. Sustainability 9(4):658. https://doi.org/10.3390/su9040658

Majumdar A, Shaw M, Sinha SK (2020) COVID-19 debunks the myth of socially sustainable supply chain: a case of the clothing industry in South Asian countries. Sustain Prod Consump 24:150–155. https://doi.org/10.1016/j.spc.2020.07.001

Mani V, Agrawal R, Sharma V (2015) Supply chain social sustainability: a comparative case analysis in indian manufacturing industries. Procedia Soc Behav Sci 189:234–251. https://doi.org/10.1016/j.sbspro.2015.03.219

Mani V, Gunasekaran A, Delgado C (2018) Supply chain social sustainability: standard adoption practices in Portuguese manufacturing firms. Int J Prod Econ 198:149–164. https://doi.org/10.1016/j.ijpe.2018.01.032

Mathiyazhagan K, Mani V, Mathivathanan D, Rajak S (2023) Evaluation of antecedents to social sustainability practices in multi-tier Indian automotive manufacturing firms. Int J Prod Res 61(14):4786–4807. https://doi.org/10.1080/00207543.2021.1938276

McMaster M, Nettleton C, Tom C, Xu B, Cao C, Qiao P (2020) Risk management: rethinking fashion supply chain management for multinational corporations in light of the COVID-19 outbreak. J Risk Financ Manage 13(8):173. https://doi.org/10.3390/jrfm13080173

Mesquita PL, Missimer M (2021) Social sustainability work in product development organizations: an empirical study of three Sweden-based companies. Sustainability 13(4):1986. https://doi.org/10.3390/su13041986

Mia R, Ahmed T, Tanjim MN, Waqar MA, Hasan MM, Das A, et al (2020) An extensive analysis of the health hazards for RMG workers in apparel sector of Bangladesh. J Textile Eng Fash Technol 6(4). https://doi.org/10.15406/jteft.2020.06.00242

Narula R (2019) Enforcing higher labor standards within developing country value chains: consequences for MNEs and informal actors in a dual economy. J Int Bus Stud 50(9):1622–1635. https://doi.org/10.1057/s41267-019-00265-1

Nath SD, Eweje G (2021) Inside the multi-tier supply firm: exploring responses to institutional pressures and challenges for sustainable supply management. Int J Oper Prod Manag 41(6):908–941. https://doi.org/10.1108/IJOPM-09-2020-0651

Oelze N, Gruchmann T, Brandenburg M (2020) Motivating factors for implementing apparel certification schemes—a sustainable supply chain management perspective. Sustainability 12(12):4823. https://doi.org/10.3390/su12124823

Perry P, Wood S, Fernie J (2015) Corporate social responsibility in garment sourcing networks: factory management perspectives on ethical trade in Sri Lanka. J Bus Ethics 130(3):737–752. https://doi.org/10.1007/s10551-014-2252-2

Rahman KM, Chowdhury EH (2020) Growth trajectory and developmental impact of ready-made garments industry in Bangladesh. In: Bangladesh's economic and social progress, pp 267–297. Springer Singapore. https://doi.org/10.1007/978-981-15-1683-2_9

Rahman S (2022) Demoralizing impacts of the COVID-19 on the Bangladesh Ready-Made Garment (RMG) supply chain, pp 191–215. https://doi.org/10.1007/978-3-030-93228-2_9

Rahman S, Rahman KM (2020) Multi-actor initiatives after Rana Plaza: factory managers' views. Dev Chang 51(5):1331–1359. https://doi.org/10.1111/dech.12572

Reza N, Du Plessis JJ (2022) The garment industry in Bangladesh, corporate social responsibility of multinational corporations, and the impact of COVID-19. Asian J Law Soc 9(2):255–285. https://doi.org/10.1017/als.2022.9

Rimi Khan HR (2022) Fashion in 'crisis': consumer activism and brand (ir) responsibility in lockdown, 1st edn. Taylorfrancis

Robinson OC (2014) Sampling in interview-based qualitative research: a theoretical and practical guide. Qual Res Psychol 11(1):25–41. https://doi.org/10.1080/14780887.2013.801543

Sakamoto M, Begum S, Ahmed T (2020) Vulnerabilities to COVID-19 in Bangladesh and a reconsideration of sustainable development goals. Sustainability 12(13):5296. https://doi.org/10.3390/su12135296

Sen S, Antara N, Sen S, Chowdhury S (2020) The apparel workers are in the highest vul-nerability due to COVID-19: a study on the Bangladesh Apparel Industry. Asia Pac J Multidisc Res 8(3):1–7

Shen L-Y, Jorge Ochoa J, Shah MN, Zhang X (2011) The application of urban sustainability indicators—a comparison between various practices. Habitat Int 35(1):17–29. https://doi.org/10.1016/j.habitatint.2010.03.006

Siddiqui J, Uddin S (2016) Human rights disasters, corporate accountability and the state. Account Audit Account J 29(4):679–704. https://doi.org/10.1108/AAAJ-07-2015-2140

Sim MR (2020) The COVID-19 pandemic: major risks to healthcare and other workers on the front line. Occup Environ Med 77(5):281–282. https://doi.org/10.1136/oemed-2020-106567

Stamm I (2023) Ecosocial work and services for unemployed people: the challenge to integrate environmental and social sustainability. Nordic Soc Work Res 13(1):134–147. https://doi.org/10.1080/2156857X.2021.1975154

Su J, Hodges NN, Wu H, Iqbal MA (2022) Coping with the COVID-19 pandemic: evidence from the apparel industry in Bangladesh and China. J Fash Market Manage Int J 1–19. https://doi.org/10.1108/JFMM-08-2021-0213

Sudusinghe JI, Seuring S (2020) Social sustainability empowering the economic sustainability in the global apparel supply chain. Sustainability 12(7):2595. https://doi.org/10.3390/su12072595

Syed RF (2024) Labor standards, labor policy, and compliance mechanism: a case study in Bangladesh. Labor History 65(2):256–272. https://doi.org/10.1080/0023656X.2023.2272124

Terwindt C, Armstrong A (2019) Oversight and accountability in the social auditing industry: the role of social compliance initiatives. Int Labour Rev 158(2):245–272. https://doi.org/10.1111/ilr.12143

Tommasetti A, Mussari R, Maione G, Sorrentino D (2020) Sustainability accounting and reporting in the public sector: towards public value co-creation? Sustainability 12(5):1909. https://doi.org/10.3390/su12051909

Ullah Z, Álvarez-Otero S, Sulaiman MABA, Sial MS, Ahmad N, Scholz M, Omhand K (2021) Achieving organizational social sustainability through electronic performance appraisal systems: the moderating influence of transformational leadership. Sustainability 13(10):5611. https://doi.org/10.3390/su13105611

Voronov M, Weber K (2020) People, actors, and the humanizing of institutional theory. J Manage Stud 57(4):873–884. https://doi.org/10.1111/joms.12559

Weingaertner C, Moberg Å (2014) Exploring social sustainability: learning from perspectives on urban development and companies and products. Sustain Dev 22(2):122–133. https://doi.org/10.1002/sd.536

# Half a Decade of the *Rohingya* Refugee Crisis: The Threats to Peace and Sustainability in Bangladesh

**Md. Bakebillah and Mokbul M. Ahmad**

**Abstract** On August 24, 2023, the Rohingya refugees marked six years since entering Bangladesh after facing a renewed cycle of genocide in the Rakhine state by the Myanmar military government. However, there is little hope for them to return to their homeland. Additionally, numerous issues such as funding crises, effects on the local community e.g., economic and social dilemmas, communal conflicts, security concerns, turf wars and environmental impacts are increasing threats to peace and sustainability. This chapter investigates the causes and consequences of rising threats to peace and sustainability for host communities in Cox's Bazar, Bangladesh due to the Rohingya refugee crisis. Moreover, it discusses the possible way forwards. The study systematically reviewed existing literature based on Scopus and Google Scholar online databases and finally examined 14 research articles and book chapters. Moreover, policy documents, media contents, and online sources are used for it. Peace and sustainability issues i.e., social, environmental and economic crises are analyzed thematically. The results illustrate that refugee crisis is a major concern that could contribute to threats to peace, and sustainability in the southeastern region, where many indigenous communities live with limited resources while the ecosystem is already under stress. International support, alongside the Myanmar government, is crucial for ensuring peaceful repatriation and livelihoods for the Rohingya people. If these challenges are not addressed soon, the cascading effects could be much more severe than they appear at present.

Md. Bakebillah (✉)
Climate Change and Sustainable Development (CCSD), School of Environment, Resources and Development (SERD), Asian Institute of Technology, Pathumthani 12120, Thailand
e-mail: st124263@ait.ac.th

Department of Folklore, Faculty of Social Science, Jatiya Kabi Kazi Nazrul Islam University, 2224 Trishal, Mymensingh, Bangladesh

M. M. Ahmad
Development and Sustainability, Asian Institute of Technology, Pathumthani 12120, Thailand
e-mail: morshed@ait.ac.th

**Keywords** Rohingya refugees · Bangladesh · Indigenous communities · Peace · Conflicts · Sustainability

# 1  Background

The Rohingya refugees completed half a decade in southeastern Bangladesh in 2023 after facing genocide in the Rakhine state, also known as Arakan, by the military junta government in Myanmar. They are now living in 34 refugee camps in Cox's Bazar district (Fig. 1). However, there is little hope for their return to Myanmar as fundamental issues such as citizenship and security remain unresolved, despite Dhaka and Naypyitaw's desire for repatriation (International Crisis Group 2023a). Additionally, numerous problems, including economic crises, conflicts, security concerns, and environmental impacts, pose significant challenges to peace and sustainable development in Bangladesh (Rahman and Siddiqi 2023).

**Fig. 1**  The location of Rohingya refugees in Cox's Bazar, Bangladesh (*Source* MDM 2021)

The Rohingya crisis began when the government of Myanmar denied citizenship to the Rohingya people in 1992 (Rhoads 2023) and violated their human rights, which the international community largely ignored (Ullah and Chattoraj 2023). The government of Myanmar declared that the Rohingya people are illegal migrants who settled there during the British colonial period for activities such as agriculture, fishing, or labor, and that they were never permanent residents of Burma/Myanmar. However, historical evidence shows that the Rohingya community has a legacy dating back to the eighth century when Arabs first arrived for trade (Uddin 2020). Moreover, Rohingya community has a centuries-long history of citizenship and active political participation with at least 22 elected members of parliament (MPs) in Myanmar between 1936 and 1990 and Rashid, a good example, a Muslim student leader and Vice-President of the Rangoon University Central Students' Union in 1936 (RUCSU) (Uddin 2020). It is worthy to mention that Aung San, father of Aung San Suu Kyi was the Secretary-General in the same committee who (Encyclopedia Britannica 2024; Uddin 2020).

The Rohingya community first fled to Bangladesh in 1978, with recent groups arriving between 2016 and 2018 (Kudrat-E-Khuda 2020; Ullah and Chattoraj 2023) while Table 1 demonstrates only the registered figures.

However, recent reports mentioned that this number reached 1.2 million by August 2022 (Prothom Alo 2023a). It may extend to around 1.6 million (Alam 2019), with approximately 30,000 Rohingyas added by new births every year (Prothom Alo 2023a). These increasing number of refugees are located mainly in several places and camps in Cox's Bazar (Fig. 2) while many are attempting to escape from camps and fleeing to various parts of Bangladesh as well as other countries.

The Rohingya refugees hosted by other countries too such as Malaysia, Saudi Arabia, Pakistan and others (Fig. 3), although a small number compared to Bangladesh. Recently, around 45,000 people are waiting near the Naf River border to enter Bangladesh because of fighting between the state army and rebel groups (Reuters 2024). However, as an overpopulated country with limited resources, the massive Rohingya population has become a burden, leading to economic, social, and environmental crises that ultimately pose threats to peace and sustainability.

In the beginning, the Rohingya issue was a global concern. However, over time, many crises have emerged globally, such as COVID-19, the Ukraine war, and the crisis in Gaza. Consequently, the global community's focus has shifted frequently (International Crisis Group 2023a; Rahman 2023). It is noteworthy that from the onset of the Rohingya crisis, both international and national media have been less focused on understanding the real facts (Ullah and Chattoraj 2023). In addition, the Bangladesh government has recently restricted the frequent movement and

| | Years | Estimated number |
|---|---|---|
| **Table 1** An estimated number of the Rohingya influx in Bangladesh (UNDP 2022) | 1978–1990 | 240,000 |
| | 2016–2018 | 621,000 |

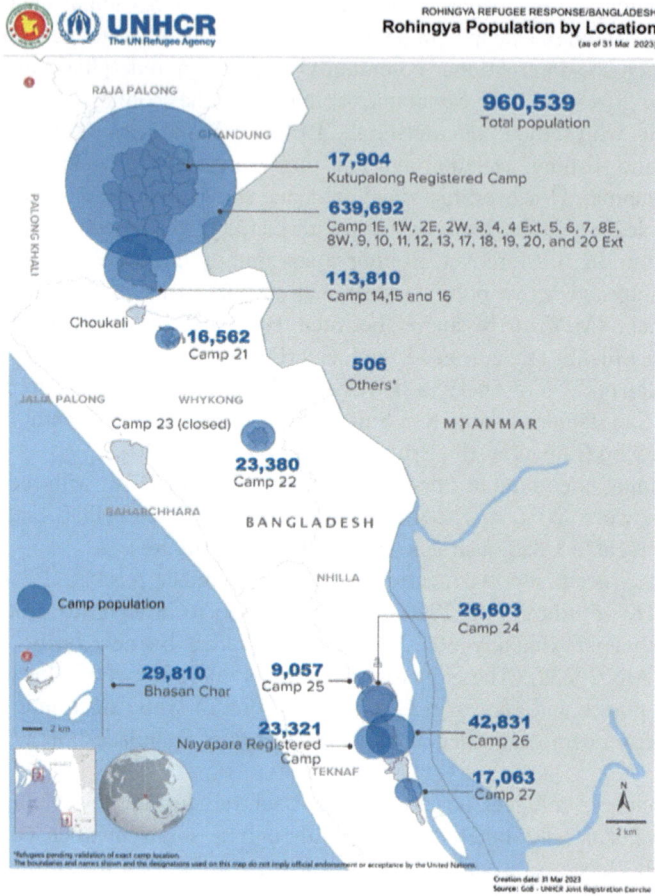

**Fig. 2** The location and Rohingya refugees living in the Cox's Bazar, Bangladesh. (UNHCR 2023a, b, c)

work opportunities for refugees outside the camps. Policymakers claim that the shortage of international funding and food supply, coupled with the host country's restrictions, may exacerbate the crisis. The Bangladeshi government needs more participation to resolve the issue. It can be argued that this situation will put more pressure on the host country from humanitarian aspects, necessitating much more aid to cope with the consequences.

There is no doubt that the Rohingya crisis is a significant humanitarian issue, and the Rohingyas have full rights to livelihood and other human rights. However, Bangladesh is one of the most densely populated countries in the world, with limited resources in a small area of 147,570 km$^2$. Moreover, the areas of the camps were originally covered by forests that are crucial for both human and biodiversity.

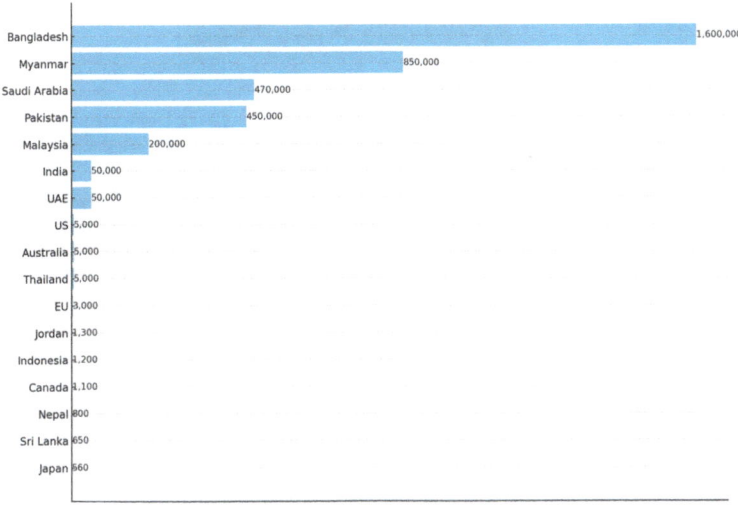

**Fig. 3** Number of Rohingya refugees hosted by different countries (Alam 2019)

Given the sensitivity of the issue for both humanitarian aspects for the Rohingya and the Bangladeshis, researchers have carefully investigated and addressed the issue. However, very few studies focused on the issue while showing a comprehensive picture in the host communities and regions in the perspective of peace and sustainability framework.

Moreover, related literature has exposed that problems have increased in several ways. For example, two repatriation initiatives, on November 15, 2018, and August 22, 2019, failed (Uddin 2024) as the Rohingya people were concerned about their citizenship, further genocidal oppression, and future sustainability, which was logical. Additionally, recent news reported that thousands of Rohingya men are being forced to join the ongoing battle between the state army and rebel groups in Myanmar (Ahmed 2024). Additionally, hundreds of Myanmar's state army soldiers have recently fled to Bangladesh due to attacks by rebel groups in the Rakhine state (Ahmed 2024). This current situation reflects that there is little hope for the Rohingya people returning to Myanmar, which may exacerbate the crisis in the refugee inhabited areas of Bangladesh.

This chapter focuses on Rohingya crisis in Cox's Bazar and how it affects and threatens peace and sustainability for the local community (Fig. 4). To address the question, the study investigates the causes and effects of social, environmental, and economic issues due to the refugee burden in Cox's Bazar, Bangladesh (Ahmad and Nusrat 2023). Additionally, the study indicates potential ways forwards for the future to mitigate the crisis, which could be useful for policymakers as well as for further research.

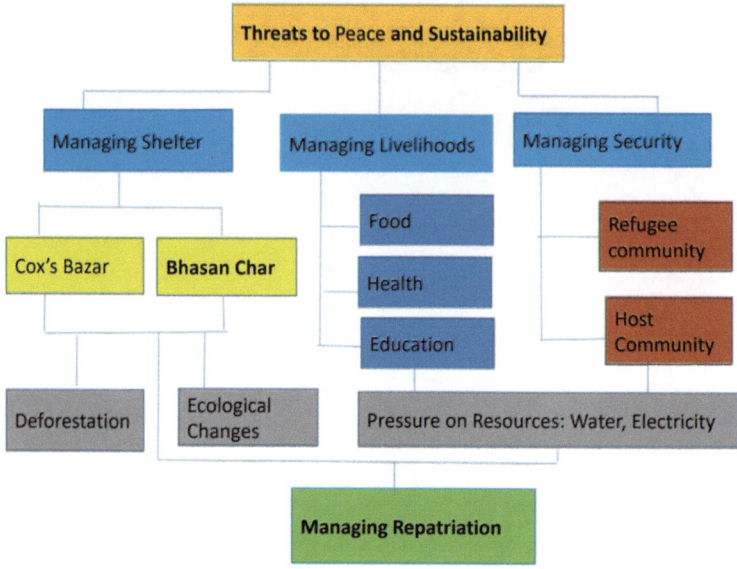

**Fig. 4** A conceptual framework for the study encompassing threats to peace and sustainability as well as a way forward for repatriation

## 2    Methods

This study analyzes the existing literature through a systematic review using the PRISMA chart which is followed into four steps (i) identification (ii) screening (iii) eligibility, and (iv) inclusion (Moher et al. 2015). Different studies like qualitative, quantitative and mixed-methods research were reviewed through maintaining a proper strategy. Moreover, media contents, policy documents and online sources were used, and data were analyzed thematically.

### 2.1    *Identification*

In terms of the exploration of literature, we used two prominent online databases: Scopus and Google Scholar. The literature survey strategy involved using systematic keywords such as "Rohingya Refugee/ Rohingya/ Rohingya people," "Conflicts/ Crisis," "Peace threats/ threat to peace," "Sustainability/ Sustainable development," "Economic, social and Environmental stress/pressure," and "Climate change" in various dimensions. The search was aligned with the study's objectives and limited to the geographical boundary of Bangladesh. Using these keywords, the databases detected 461 articles (280 from Scopus and 181 from Google Scholar).

## 2.2   Screening

From the identified articles (n = 461), the study scanned all the titles of articles and excluded 264 articles that do not match with objectives of the study and remained 197 articles. Then 197 abstracts screened and excluded 132 while 14 we detected as duplicated that we also deleted from the review list, and it reached 51 articles.

## 2.3   Eligibility

After screening the title and abstract, 51 articles we selected for the eligibility check. Only Bangladesh-based publications were allowed for selection and other regions were extracted between April 2018 and April 2024. Different research domains like social science, environmental science and art-humanities are included and engineering, earth science and others were excluded. Scopus and Google Scholar databases searched for selecting journal articles and book chapters while others like review articles were excluded. Only peer-reviewed and English language-based articles were accepted where predatory journals and non-English research were excluded. Overall, the study considered peace and sustainability threat issues i.e., social, environmental, and economic stress due to the Rohingya refugee crisis in Coz's Bazar in Bangladesh.

## 2.4   Inclusion

A total of 51 articles were initially considered after applying the inclusion and exclusion criteria. Subsequently 37 articles were excluded for having little relation to the study focus. Therefore, the study ultimately reviewed the remaining articles (n = 14) (Fig. 5).

## 3   Results

As the largest influx of Rohingya people occurred in 2017, the chapter focused on publications between 2018 and April 2024. Although 461 articles were initially identified using keyword searches, only 14 articles were selected after several stages of screening. The Rohingya issue is a priority area for research, but most studies have focused on the perspectives of the Rohingya people, with limited research available on the host community or aspects related to Bangladesh. Additionally, predatory journal articles were avoided in this chapter. Figure 6

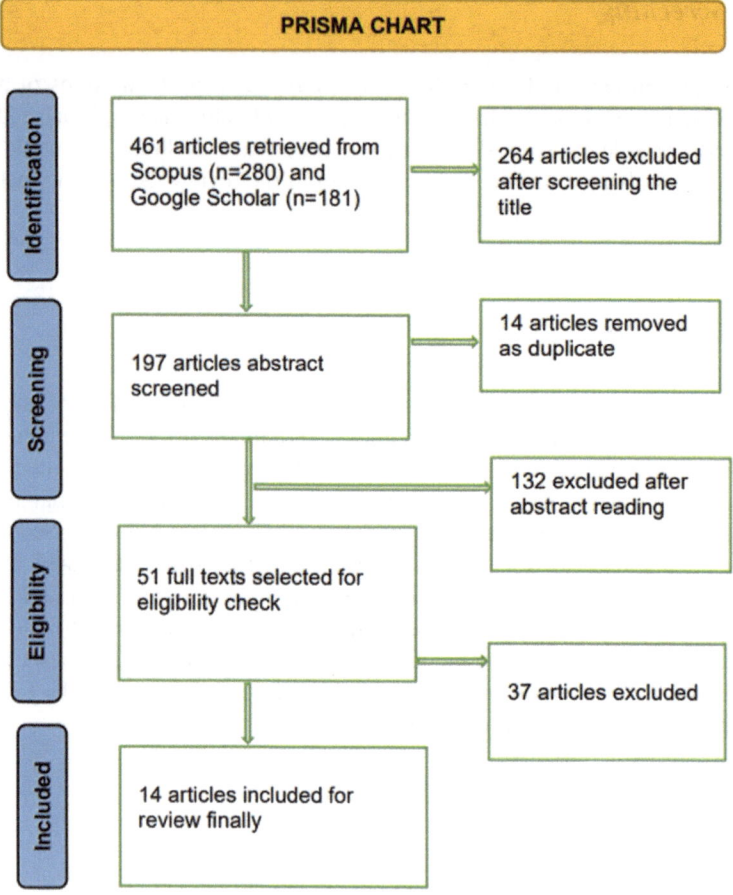

**Fig. 5** PRISMA follows chart of the systematic review of the study

shows that articles published in 2022 constitute the largest portion of the review. Additionally, articles from the years 2020, 2023, and 2024 also contributed significantly, demonstrating a consistent interest in the Rohingya issue over the years.

Peace and sustainability are integrated issues (Sharifi et al. 2021). Environmental damage can lead to economic losses, which in turn can cause social tensions and finally threaten peace. However, a review of the summary in Table 3 reveals that social insecurity is the primary concern, with 10 papers discussing a wide range of issues including conflicts, security concerns, terrorism, drug trafficking, human trafficking, disparity, frustration, neighborhood violence, hostility, and overall threats to peace.

Additionally, environmental damage is another concern, encompassing deforestation, land degradation, ecosystem and biodiversity loss (Hassan et al.

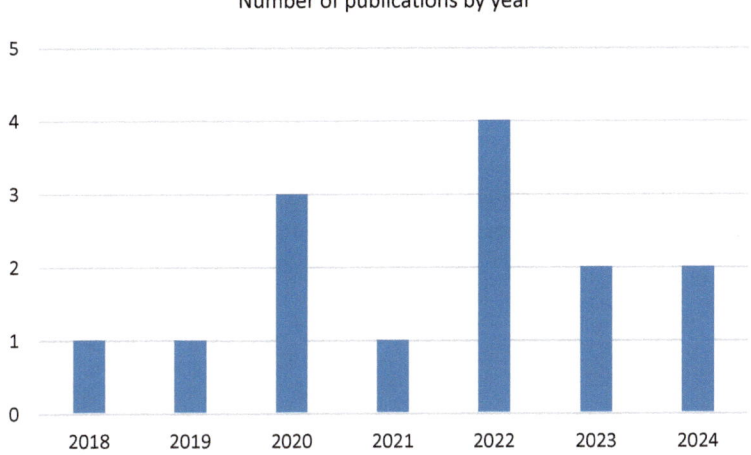

**Fig. 6** Reviewed articles based on publication year from 2018 to April 2024

2018), all of which contribute to climate change (Neef et al. 2023). Although economic loss is discussed less frequently (Alam et al. 2022) compared to social and environmental issues (Fig. 7), it remains a significant concern, particularly as Bangladesh is currently facing a severe financial crisis. It is also important to note that donors are diverting their aid elsewhere due to various global wars and conflicts.

## 4  Discussion: Threats to Peace and Sustainability

The UN member countries agreed in 2015 to achieve sustainable development goals (SDGs) for peace and overall prosperity for human beings and the planet (UN 2015). Moreover, peace researchers argue for discovering both causes and consequences war lead to solutions while Asia–Pacific perspectives primarily focus on the problems of interstate and intrastate war (Peou 2010). However, the Rohingya crisis is not only an intrastate and interstate war but is also broadly related to economic, social, and environmental sustainability issues (Singh et. al., 2009) that ultimately threaten peace. Therefore, some scholars debate that there is no fundamental difference between peace and sustainable development, asserting that sustainable development is essential for peace and vice versa (Sharifi et al. 2021). The study establishes a nexus between peace and sustainability issues, suggesting that sustainability indicators can serve as a conduit for achieving peace.

**Table 3** A summary of the reviewed articles

| References | Themes | Key findings |
|---|---|---|
| Ahmed and Sabastini (2024) | Environmental stress and conflict | • Significant stress on livestock<br>• Loss of agroforestry<br>• Increasing food insecurity<br>• Growing sense of deprivation among host communities |
| Kamruzzaman et al. (2024) | Conflicts, crimes, environmental and livelihoods stress | • Local people towards Rohingya refugees have turned into frustration and hostility over time<br>• Unlawful activities<br>• Loss of forests and livelihood opportunities<br>• Disparity in humanitarian support<br>• Rising living costs and growing tension |
| Neef et al. (2023) | Climatic hazards, conflict, and displacement | • Rising climate disaster risks, political conflicts, and forced displacement |
| Sakib (2023) | Security concern and crimes | • Drug trafficking<br>• Terrorism and arms smuggling<br>• Human trafficking, and fraud |
| Siddiqi (2022) | Conflicts, financial and environmental challenges, and crimes | • Declining sympathy of in the host communities<br>• Facing financial challenges<br>• Environmental degradation<br>• Drug trafficking<br>• Neighborhood violence<br>• Security threats |
| Sajib et al. (2022) | Rohingya crisis | • Impacts on farmland, local environment, ecosystems, and biodiversity<br>• Increasing pressure on resources<br>• Socio-economic tension between the local and Rohingya communities |
| Hassan et al. (2023) | Environmental stress | • 600 acres of protected forests were affected by 2017 in Teknaf<br>• 7500 acres of non-protected forests are affected |
| Alam et al. (2022) | Economic stress | • Prices increased by 8%, with a 7–36% increase for protein and vegetables in Ukhia, in Cox's Bazar district |

(continued)

**Table 3** (continued)

| References | Themes | Key findings |
|---|---|---|
| Ahmed et al. (2021) | Environmental stress and conflicts | • Severe deforestation<br>• Habitat loss<br>• Impacts on internal climate variability<br>• Increased conflict threats |
| Hasan et al. (2020) | Environmental stress | • Significant forest degradation between 2017 and 2019<br>• A loss of 3130 ha of forest<br>• The study forecasts that by 2027, this loss could reach 5115 ha |
| Bari (2020) | Conflicts | • Serious tensions have arisen due to clashes between the local community and Rohingya people over limited resources |
| Kudrat-E-Khuda (2020) | Economic, social and environmental stress | • Significant economic losses, social impacts, and environmental damages |
| Islam (2019) | Conflicts | • Concern for communal violence between different religious and ethnic groups |
| Hassan et al. (2018) | Environmental stress | • Significant forest land (1219 ha) loss by 2018<br>• Land degradation, biodiversity loss<br>• Destruction of elephant corridors<br>• Elephants being killed by Rohingya people |

## 4.1 Social Dilemma

The study identified several issues that threaten peace at the regional level and may extend beyond national boundaries such as, social problems are increasing significantly (Kudrat-E-Khuda 2020), including conflicts (Ahmed and Sabastini 2024; Kamruzzaman et al. 2024; Neef et al. 2023; Siddiqi 2022; Bari 2020), security issues (Sakib 2023; Siddiqi 2022), communal threats (Sajib et al. 2022; Islam 2019), drug trafficking, human trafficking, domestic violence (Sakib 2023; Siddiqi 2022); , and early marriage (MSF 2023).

As many ethnic communities, including Bengalis, live in Cox's Bazar, communal conflicts and tensions are rising due to overpopulation, resource scarcity, and heightened crime (Crabtree 2010). A major communal conflict occurred in Teknaf, Ukhia, and Ramu on September 29–30, 2012, when local Buddhist members were attacked by other community members based on religious provocations while

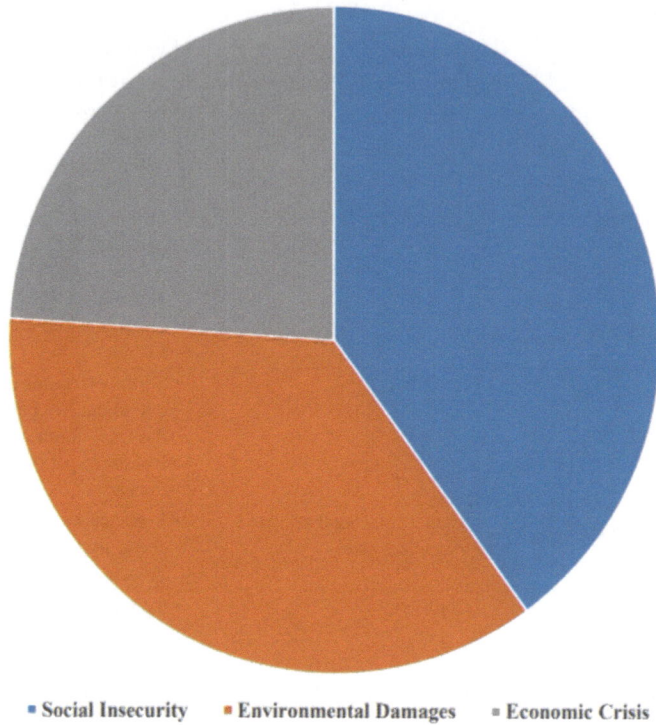

■ Social Insecurity    ■ Environmental Damages    ■ Economic Crisis

**Fig. 7** Ration of reviewed articles on sustainability issues

houses, temples (prayer halls), and other assets were destroyed, and 316 people were arrested by police and many Rohingyas who had settled there after previous influxes were blamed for involvement in the attacks (Banglanews24 2023).

Bangladesh is a Muslim-majority country with various ethnic communities living throughout the country. At times, local movements have demanded that the Rohingyas return to Myanmar (Banglanews24 2023). It is evident that local people are dissatisfied with the settlement of Rohingya refugees (Kamruzzaman et al. 2024). Over time, tension has increased, manifesting in new forms of communal violence between different religious and ethnic groups (Islam 2019. The initial sympathy of local people toward Rohingya refugees has gradually turned into frustration and hostility (Kamruzzaman et al. 2024; Siddiqi 2022).

In recent years, security in the refugee camps and surrounding areas in Cox's Bazar has deteriorated significantly, with at least 50 murders reported in the camps in the first half of 2023, although the real figure may be higher. The main reason is that armed and criminal groups have established a strong foothold inside the camps (International Crisis Group 2023b). These groups are involved in Yaba smuggling, a drug trade from Myanmar that crosses the Naf River into

Bangladesh. Many Rohingyas earn money for their livelihoods through such activities, with some stating that they carry *Yaba* to survive (Alsaafin 2018).

National and international drug smugglers use Rohingya people as pawns in this crime business. Human trafficking, kidnapping, drug business, terrorism, arms smuggling, and fraud are regular occurrences (Sakib 2023; Siddiqi 2022). The Arakan Rohingya Salvation Army (ARSA) and Rohingya Solidarity Organisation (RSO) are well-known for such illegal activities, although refugees initially believed these groups would help them return to their homeland. Other armed groups, such as the Nobi Hossain group, the Munna group, and Islami Mahas, are also expanding their influence through intra-group conflicts and illegal activities. It is estimated that at least 11 groups are involved in these activities (IISS 2023; Shawon 2023).

Additionally, the International Crisis Group (2023b) reported that around 700 abductions took place by September 2023 which were significant higher compared to 2022 and since 2017 every year a largest number of Rohingya people were accused and reported by the police (Fig. 8). Data illustrated that in 2022, some refugees attempted to move via sea voyages to find shelter abroad which a 360% increase compared to previous years, with 348 individuals reported dead or missing (UNHCR 2023a, b, c). Prothom Alo, the most circulated newspaper in Bangladesh, reported on November 20, 2023, that ARSA has become a significant threat to the refugee camps. Recently, three commanders were arrested by the elite security forces RAB, as the group is accused of murder, kidnapping, trafficking, and drug smuggling, and many young Rohingyas are joining such groups. Moreover, at least 76 Rohingya killed by Rohingya terrorists by October 2023 (Fig. 9) which were double in number compared to 2022 (IISS 2023).

Recently, three Rohingya people died in a gunfight between ARSA and RSO, with 61 fights occurring in the past eight months, resulting in 74 deaths, according to police and Rohingya leaders (Prothom Alo 2023b). ARSA was once known as

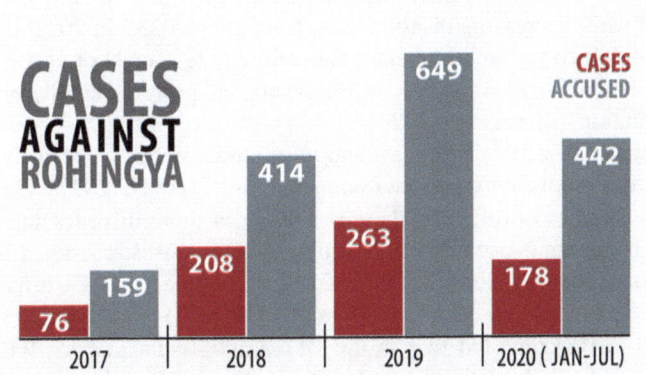

**Fig. 8** Reported several cases and accused scenario against the Rohingyas, Cox's Bazar, Bangladesh (*Source* Khan 2020)

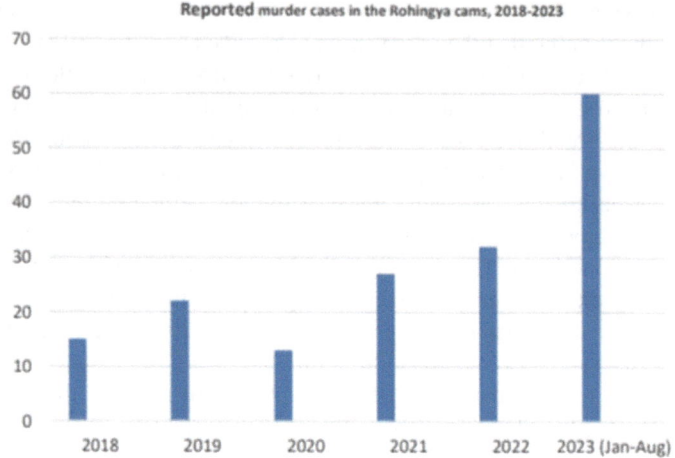

**Fig. 9** Reported murder cases according to Cox's Bazar district police (*Source* Khan 2020; Prothom Alo 2023b)

the 'night government' of the camps (IISS 2023). However, the situation is now more dangerous, as refugees are frequently attacked during the day, and even women and children are victimized (International Crisis Group 2023b). No doubt, if the crisis continues, it could disrupt communal harmony and exacerbate political conflicts in the region (Neef et al. 2023).

## 4.2 Environmental Damages

Moreover, environmental damage is a major concern in the Rohingya camp areas in Cox's Bazar (UNDP 2022; Ahmed and Sabastini 2024; Siddiqi 2022; Hassan et al. 2018) while increasing disaster risks (Neef et al. 2023). In 2022, the UN conference highlighted the environment is one of the core agenda of as the 122 member countries agreeing to manage natural resources, protect the environment, and prevent pollution (Sharifi et al. 2023).

Cox's Bazar is one of the longest and most attractive sea beaches in the world, with numerous biodiversity and environmental assets (UNDP 2022). However, due to the large number of refugees, there is a need for more infrastructure and habitats while the region is one of the most vulnerable and disaster risks in Bangladesh due to climate change (Ahmed et al. 2021) for example, in 2017, around 70% of shelters were damaged by a cyclone Mora (UNDP 2022).

The UN (2018) reported that in the Ukhia and Teknaf areas, 793 ha of forests were used for habitation and fuel, out of a total of 1502 ha of natural forests. Another study showed that significant forest degradation occurred in the Cox's Bazar and Teknaf peninsula due to Rohingya refugee pressure between 2017 and

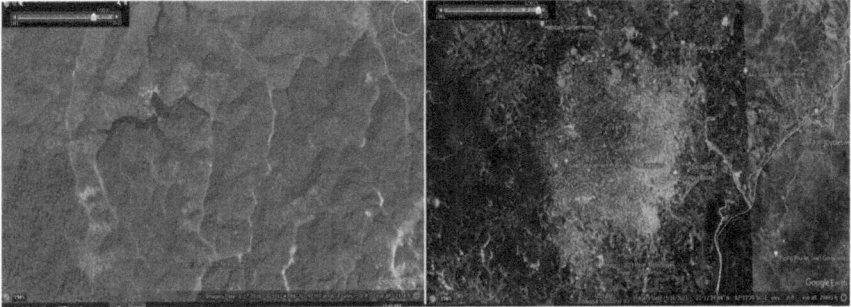

**Fig. 10** A comparison between 2016 and 2023 in Kutupalong and Balukhal Rohingya refugee camps, Cox's Bazar in Bangladesh (*Source* Google Earth)

2019, resulting in the loss of 3130 ha of forest and the study forecasted that by 2027, the loss could reach 5115 ha (Hasan et al. 2020). Additionally, research illustrated that forests were destroyed significantly (21.97%), equivalent to a loss of $5.33 million, while land, raw materials, and overall biodiversity declined by 13.58% and 14.57%, respectively (Sarkar et al. 2023). Human-wildlife conflicts are also increasing, with several reports mentioning that wildlife has been killed by the refugees for meat and other resources (UNDP 2022; Hassan et al. 2018)).

Figure 10 shows the before and after conditions of Kutupalong and Balukhali camp areas. In 2016, the area was green, but by 2023, there was little space for greenery due to overpopulation and the establishment of refugee habitats. Deforestation for residents' materials and fuel is a common issue, with 86% of people involved in the cutting and selling of wood (Rahman 2017). The overall impacts include land degradation and biodiversity loss, such as the destruction of elephant corridors and the killing of elephants by refugees. The entire ecosystem in these areas is under severe threat (Hassan et al. 2018).

## *4.3 Economic Crisis*

The economic crisis is increasing over time associated with national and international funding such as basic needs, income issues within host communities, and impacts on economic factors (Alam et al. 2022; Kudrat-E-Khuda 2020). This integrated process threatens peace and sustainability in the host community and country, acting like interconnected cells.

Future economic challenges are a major issue, as there is a need to manage the daily livelihood costs of 1.1 million registered Rohingya refugees, with many remaining unregistered. Funds for this large population are provided by numerous donor countries and international aid agencies through several initiatives, such as the "Joint Response Plan" (JRP) for the Rohingya refugees by the UN. However,

**Table 4** A comparison of funding scenario from 2017 to 2022 (*Source* Rahman 2023)

| Funding year | Need USD/million | Available/USD/million | Shortage of funding (%) |
|---|---|---|---|
| 2022 | 881.0 | 433.1 | 50.8 |
| 2021 | 943 | 678.5 | 28.1 |
| 2020 | 1058 | 629 | 40.5 |
| 2019 | 920 | 699 | 24 |
| 2018 | 951 | 655 | 31 |
| 2017 | 434.1 | 317 | 27 |

these funds have never been sufficient and have been decreasing each year, as shown in Table 4.

This scenario remains unchanged. The Joint Response Plan (JRP, 2023) appealed for USD 876 million in funding for 2023, yet only 28.9% of this target was achieved (UNHCR 2023a, b, c; International Crisis Group 2023b). The data unequivocally indicates a decline in funding, which will likely aggravate the associated impacts. Consequently, the question arises: who is covering the remaining costs? The host country, Bangladesh, is bearing the burden in various ways. To date, 1 billion BDT (over USD 90 million) has been allocated for activities such as the installation of security fences. Additionally, under the Ashrayan Project-3 (Fig. 11), Bangladesh has expended USD 380.1 million to construct the Bhasan Char shelter for 1 million Rohingya refugees (Rahman 2023).

As funding and relief are decreasing, refugees are struggling to meet basic needs such as food. Some girls are becoming involved in sex work, marrying at an early age, attempting dangerous boat journeys to migrate to Malaysia or Thailand,

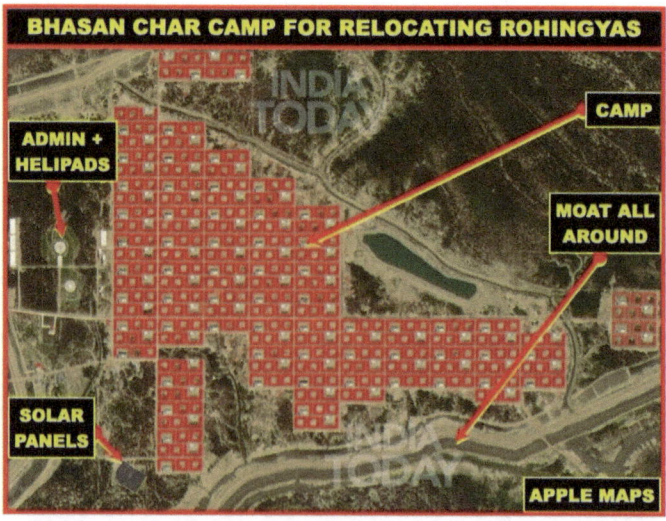

**Fig. 11** The Ashrayan Project-3 (Bhat 2020)

and borrowing money from local brokers at exorbitant interest rates (International Crisis Group 2023b).

Meanwhile, Bangladesh faces several impending economic crises. Security concerns in the Cox's Bazar and Chattogram areas are escalating, threatening the tourism sector (Raihan 2017). Numerous economic and industrial projects in these areas rely heavily on foreign investment. If the crisis continues to escalate, all investments and economic activities will be significantly impacted.

Furthermore, stress on livestock rearing, loss of agroforestry, and increasing food insecurity are critical issues as the funding crisis intensifies (Ahmed and Sabastini 2024). Food and living costs are rising significantly (Kamruzzaman et al. 2024). A study showed that overall prices increased by 8%, with protein and vegetable prices rising by 7–36% in Ukhia, a host sub-district for Rohingya refugees in Cox's Bazar district (Alam et al. 2022).

## 5  The Way Forward

The Rohingyas have been among the most oppressed and victimized people in the world over the past few decades, and the distress they have faced in the last six years seem to have exceeded all previous records. Repatriation and the confirmation of citizenship by the Myanmar government are essential to resolving this issue. As citizens of Myanmar, they deserve a peaceful and secure life. Recent attempts by Myanmar authorities to repatriate Rohingya people have failed twice because the Rohingyas demanded citizenship first (BBC News 2019). Both Myanmar and the international community have committed to meeting these demands. However, every stakeholder must work together wholeheartedly to facilitate the return of the Rohingyas.

In recent years, security issues have increased significantly. While these issues primarily affect the Rohingya people, they will eventually impact the host community as well. If terrorist groups like ARSA expand their activities and more young people join these groups (MSF 2023), it will become a major concern and threat to peace and sustainability in Bangladesh. Therefore, it is crucial to take immediate action to eliminate all terrorist groups and illegal activities such as drug trafficking, human trafficking, and kidnapping. Additionally, urgent policies and initiatives are needed to address community development and harmony (Islam 2019).

Given the dramatic destruction of ecosystems, restoration efforts must be prioritized. This can be achieved through measures such as reforestation, biodiversity preservation, and community engagement (Mahamud et al. 2022). Ultimately, all humans have the right to live in peace. In addition to ensuring basic human needs, a secure homeland must be prepared for the Rohingya people. Both short-term and long-term policies should be implemented by the host and international communities, with Myanmar taking the lead in these initiatives. If not addressed, this issue will not only impact Bangladesh but could also pose a future threat to peace in South Asia and Southeast Asia. Specifically, the study recommends way out while need to contribute by both Bangladesh, Myanmar as well as international communities.

## 5.1  Bangladesh

As peaceful repatriation is the best way to solve this problem, Bangladesh should pay more attention to how they can advance and execute through bilateral agreement with Myanmar. In this case Bangladesh can take support from a third party who is a good friend for both like China. Secondly, need to be ensured effective initiatives by international parties i.e., UN. Both these initiatives could be ensured by the Bangladesh government diplomacy.

## 5.2  Myanmar

Bangladesh can talk and take initiatives a lot. But only Myanmar can ensure a peaceful repatriation to receive their people by ensuring citizenship, dignity as well as future security and sustainability.

## 5.3  International

The Rohingya community is one of the worst victims of the world for the last four decades while insufficient funding and few other supports provided by international parties. But, as humanitarian cases, stronger initiatives like proper policy, advocacy, funding support are expected by them. Otherwise, the situation will be very tough for Bangladesh.

## 5.4  Management Repatriation

Currently a war is going on between the state army and rebel groups in Myanmar. It should ensure that there is no new enforcement to flee by the state army or rebel groups and after solving the war how to execute the repatriation within convenient time that also start to work. However, both Bangladesh, Myanmar and international parties must work together while civil society, NGOs, academia and others could commit significant impacts.

# 6   Conclusion

The findings emphasize that the Rohingya refugee crisis and its associated impacts are a significant concern, particularly in camp areas where indigenous communities live with limited resources. The threats to communal harmony and security are substantial. The presence of over a million Rohingya refugees has led to a pressing need for habitats and livelihoods, resulting in significant deforestation and biodiversity loss in the hilly regions, further degrading the ecosystem. This situation poses a severe threat to regional peace and presents numerous challenges to sustainable development. If these issues are not promptly addressed, sustainable development will be hindered, and threats to peace will escalate in Bangladesh and its neighboring countries.

# References

Alam A, Dutta I, Haque ME, Nogales R (2022) Impact of Rohingya refugees on food prices in Bangladesh: evidence from a natural experiment. World Dev 154:105873. https://doi.org/10.1016/j.worlddev.2022.105873

Alam S (2019) Infographic: top Rohingya-hosting countries. Anadolu Agency. https://www.aa.com.tr/en/asia-pacific/infographic-top-rohingya-hosting-countries/1563674

Ahmad MM, Nusrat R (2023) Connecting the dots: assessing the role of 'women, peace and security' agenda in conflict resolution and peacebuilding in climate fragile zones of the rohingya camps in bangladesh. In: Sharifi A, Simangan D, Kaneko S (eds) Bridging peace and sustainability amidst global transformations. Springer Nature Singapore, Singapore, pp 247–267. https://doi.org/10.1007/978-981-99-7572-3_13

Ahmed S, Simmons WP, Chowdhury R, Huq S (2021) The sustainability–peace nexus in crisis contexts: how the Rohingya escaped the ethnic violence in Myanmar but are trapped into environmental challenges in Bangladesh. Sustain Sci 16(4):1201–1213. https://doi.org/10.1007/s11625-021-00955-6

Ahmed K (2024) Thousands flee after Myanmar rebelsuse drones to bomb Rohingya villagers. Gurdian org https://www.theguardian.com/globaldevelopment/article/2024/aug/22/myanmar-rohingya-muslim-minority-flee-bangladesh-massacres-arakan-armyrebels

Ahmed T, Sabastini P (2024) Deforestation as a site of conflict and differentiation: the case of the Rohingya refugee influx in the Teknaf Wildlife Sanctuary. Hum Ecol. https://doi.org/10.1007/s10745-024-00480-x

Alsaafin L (2018) 'I carry Yaba to survive': Rohingya and Bangladesh's meth trade. Al Jazeera. https://www.aljazeera.com/features/2018/8/19/i-carry-yaba-to-survive-rohingya-and-bangladeshs-meth-trade

Bhat CV (2020) Bangladesh shifting Rohingyas to Bhasan Char Island that lacks basic facilities, reveal satellite images. India Today. https://www.indiatoday.in/world/story/bangladesh-shifting-rohingyas-bhasan-char-island-lacks-basic-facilities-satellite-images-1747755-2020-12-08

Banglanews24 (2013) কক্সবাজারের আলোচিত ৫ ঘটনা. Banglanews24.com. https://www.banglanews24.com/cat/news/bd/162622.details

BBC News (2019) জলবায়ু পরিবর্তনের প্রভাব. BBC News Bangla. https://www.bbc.com/bengali/news-49638483

Bari S (2020) The Rohingya refugee crisis: a time bomb waiting to explode. Social Change 50(2):285–299. https://doi.org/10.1177/0049085719901038

Crabtree K (2010) Economic challenges and coping mechanisms in protracted displacement: a case study of the Rohingya refugees in Bangladesh. J Muslim Mental Health 5(1):41–58. https://doi.org/10.1080/15564901003610073

Encyclopaedia Britannica (2024) Aung San: Myanmar nationalist. https://www.britannica.com/biography/Aung-San

Hasan ME, Zhang L, Dewan A, Guo H, Mahmood R (2020) Spatiotemporal pattern of forest degradation and loss of ecosystem function associated with Rohingya influx: a geospatial approach. Land Degrad Dev 32(13):3666–3683. https://doi.org/10.1002/ldr.3821

Hassan MM, Duveneck M, Southworth J (2023) The role of the refugee crises in driving forest cover change and fragmentation in Teknaf, Bangladesh. Ecol Inform 74:101966. https://doi.org/10.1016/j.ecoinf.2022.101966

Hassan MM, Smith AC, Walker K, Rahman MK, Southworth J (2018) Rohingya refugee crisis and forest cover change in Teknaf, Bangladesh. Remote Sens 10(5):689. https://doi.org/10.3390/rs10050689

IISS (2023) Competing armed groups pose new threat to Rohingya in Bangladesh. https://myanmar.iiss.org/analysis/rohingya

Islam MS (2019) Communal peace in Bangladesh after Rohingya crisis: an assessment of different approaches to community development. J Prev Interv Community 48(3):256–271. https://doi.org/10.1080/10852352.2019.1625604

International Crisis Group (2023a) Crisis mounts for Rohingya refugees in Bangladesh. https://www.crisisgroup.org/asia/south-asia/bangladesh/355-crisis-mounts-rohingya-refugees-bangladesh#:~:text=Fighting%20between%20the%20once%2Ddominant,increased%20nearly%20fourfold%20in%202023

International Crisis Group (2023b) Rohingya refugees in Bangladesh: limiting the damage of a protracted crisis. https://www.crisisgroup.org/asia/south-east-asia/myanmar-bangladesh/rohingya-refugees-bangladesh-limiting-damage-protracted

JRP (2023) Rohingya response. https://rohingyaresponse.org/project/2023-jrp/

Kamruzzaman P, Siddiqi B, Ahmed K (2024) Navigating the shift in Bangladeshi host community's perceptions towards the Rohingya refugees: a declining sympathy. Front Sociol 9. https://doi.org/10.3389/fsoc.2024.1346011

Khan MJ (2020) Refugee camps in Cox's Bazar: Rohingyas tangled up in crimes. The Daily Star. https://www.thedailystar.net/frontpage/news/refugee-camps-coxs-bazar-rohingyas-tangled-crimes-1951517

Kudrat-E-Khuda (2020) The impacts and challenges to host country Bangladesh due to sheltering the Rohingya refugees. Cogn Soc Sci 6(1). https://doi.org/10.1080/23311886.2020.1770943

Mahamud R, Jalal R, Ritu S, Donegan F, Arif TA, Kumar MF, Henry M (2022) Restoring degraded land in Rohingya refugee camps in Cox's Bazar, Bangladesh. Food and Agriculture Organization of the United Nations Rome. https://openknowledge.fao.org/server/api/core/bitstreams/64716592-d871-46a8-81d9-4759f72dd951/content

MDM (2021) Rohingya refugee community support project. 国際協力 NGO 世界の医療団.国際協力 NGO 世界の医療団. https://www.mdm.or.jp/project-en/21364/

Moher D, Shamseer L, Clarke M, Ghersi D, Liberati A, Petticrew M, Shekelle P, Stewart LA (2015) Preferred reporting items for systematic review and meta-analysis protocols (PRISMA-P) 2015 statement. Syst Rev 4(1). https://doi.org/10.1186/2046-4053-4-1

MSF (2023) Rohingya youth trapped in violence and despair in Cox's Bazar. [Video]. YouTube. https://www.msf.org/rohingya-youth-trapped-violence-and-despair-coxs-bazar

Neef K, Jones E, Marlowe J (2023) The conflict, climate change, and displacement nexus revisited: the protracted Rohingya refugee crisis in Bangladesh. J Peacebuild Dev 18(3):231–247. https://doi.org/10.1177/15423166231190040

Prothom Alo (2023a) চার বছরে রোহিঙ্গার সংখ্যা বেড়ে ১২ লাখ. Prothomalo.com. https://www.prothomalo.com/bangladesh/uj0cq5yw1k

Prothom Alo (2023b) আশ্রয়শিবিরে আরসা-আরএসওর গোলাগুলি, ৩ রোহিঙ্গা নিহত. Prothomalo. https://www.prothomalo.com/bangladesh/district/ntlykwr1d2

Rahman DSA (2023) Rohingya crisis: the picture of fund crisis and its impacts on Bangladesh. Modern Diplomacy. https://moderndiplomacy.eu/2023/01/12/rohingya-crisis-the-picture-of-fund-crisis-and-its-impacts-on-bangladesh/

Rahman H, Siddiqi B (2023) The role of media and social cohesion between host and the Rohingyas in Cox's Bazar. In: World sustainability series, pp 115–135. https://doi.org/10.1007/978-981-19-7295-9_8

Rahman MZ (2017) Livelihoods of Rohingyas and their impacts on deforestation. In: Springer eBooks, pp 113–125. https://doi.org/10.1007/978-981-10-5475-4_9

Raihan S (2017) বাংলাদেশের অর্থনীতিতে বিপদ কোথায়? Prothomalo.com. https://www.prothomalo.com/opinion/column

Reuters (2024) Myanmar's Rohingya in the crosshairs as fighting escalates in Rakhine. Deccan Herald. https://www.deccanherald.com/world/myanmars-rohingya-in-the-crosshairs-as-fighting-escalates-in-rakhine-3040136

Rhoads EL (2023) Citizenship denied, deferred and assumed: alegal history of racialized citizenship in Myanmar. Citizsh Stud 27(1):38–58. https://doi.org/10.1080/13621025.2022.2137468

Sajib SMSA, Islam SAMZ, Sohad MKN (2022) Rohingya influx and socio-environmental crisis in southeastern Bangladesh. Int J Commun Soc Dev 4(1):89–103. https://doi.org/10.1177/25166026211067604

Sakib ABMN (2023) Rohingya refugee crisis: emerging threats to Bangladesh as a host country? J Asian Afr Stud. https://doi.org/10.1177/00219096231192324

Sarkar SK, Saroar M, Chakraborty T (2023) Navigating nature's toll: assessing the ecological impact of the refugee crisis in Cox's Bazar, Bangladesh. Heliyon 9(7):e18255. https://doi.org/10.1016/j.heliyon.2023.e18255

Sharifi A, Simangan D, Kaneko S (2023) Integrated approaches to peace and sustainability. In: World sustainability series. Springer. https://doi.org/10.1007/978-981-19-7295-9

Sharifi A, Simangan D, Kaneko S, Virji H (2021) The sustainability-peace nexus: why is it important? Sustain Sci 16(4):1073–1077. https://doi.org/10.1007/s11625-021-00986-z

Shawon AA (2023) How Rohingya camps are turning into crime zones. Dhaka Tribune. https://www.dhakatribune.com/bangladesh/rohingya-crisis/306260/how-rohingya-camps-are-turning-into-crime-zones

Siddiqi B (2022) Challenges and dilemmas of social cohesion between the Rohingya and host communities in Bangladesh. Front Human Dyn. https://doi.org/10.3389/fhumd.2022.944601

Singh RK, Murty H, Gupta S, Dikshit AK (2009) An overview of sustainability assessment methodologies. Ecol Ind 9(2):189–212. https://doi.org/10.1016/j.ecolind.2008.05.011

UNDP (2022) Report on environmental impact of Rohingya influx. https://www.undp.org/bangladesh/publications/report-environmental-impact-rohingya-influx

UNHCR (2023) Rohingya population by location. https://data.unhcr.org/en/documents/details/100045

UNHCR (2023) Six years since the Rohingya refugee influx in Bangladesh. https://reliefweb.int/report/bangladesh/six-years-rohingya-refugee-influx-bangladesh-unhcr-appeals-sustained-support-and-solutions-enbn

UN (2015) The 17 goals: sustainable development. https://sdgs.un.org/goals

UN (2018) Report on Environmental Impact of Rohingya Influx. Dhaka, Bangladesh,Published by UNDP Bangladesh and UN WOMEN

Uddin N (2020) The Rohingya: an ethnography of "subhuman" life, 1st edn. Oxford University Press

Uddin N (2024) Understanding 'refugee resettlement' from below: decoding the Rohingya refugees' lived experience in Bangladesh. World Dev 181(2024):106654. https://doi.org/10.1016/j.worlddev.2024.106654

Ullah AA, Chattoraj D (2023) The unheard stories of the Rohingyas: ethnicity, diversity and media, 1st ed. Bristol University Press. https://doi.org/10.2307/jj.5186775

UNHCR (2023) Protection at sea in South-East Asia-2022 in review. https://data.unhcr.org/en/documents/download/98170

# Social Entrepreneurship for Returning Filipino Indigenous Migrant Workers

Anna Joceline Dizon Ituriaga

**Abstract** Return migration has become a significant socioeconomic issue, affecting both returnees and their home countries. Although previous research has largely focused on the general experiences of return migrant workers and their entrepreneurial reintegration, few studies have examined the reintegration of Indigenous migrant workers and the potential of social entrepreneurship to address the Philippines' development challenges of high poverty and inequality. To fill this gap, I investigate the sustainability of social entrepreneurship as a livelihood program for Indigenous Peoples. My study is organized around three main themes: the reintegration of migrant workers through entrepreneurship, the role of social entrepreneurship, and the specific reintegration of Indigenous migrants through social enterprises. My findings suggest that social entrepreneurship could be a viable alternative for the economic reintegration of returning Indigenous migrant workers.

**Keywords** Social enterprises · Reintegration · Indigenous peoples · Overseas Filipino workers (OFWs)

## 1 Introduction

The Philippines is regarded as one of the world's top labor-exporting countries, with foreign personal remittances totaling $33.46 billion in 2019 (Arceo 2020). In the early 2000s, scholars began to pay attention to return migration in the Philippines (Banta 2020; Saguin 2020; Wickramasekara 2019). These scholars concentrated mostly on the overall nature of OFW reintegration. However, there is little research available on the reintegration of Overseas Filipino Workers (OFWs) in areas outside of Metro Manila, particularly the indigenous migrant workers. In this chapter, I investigate the Philippines' reintegration programs and the sustainability of social enterprises in their reintegration into Indigenous communities. One reason for this is that social enterprises serve the disadvantaged population. Therefore, there is a

A. J. D. Ituriaga (✉)
National Chengchi University, Taipei, Taiwan
e-mail: 168250@o365.tku.edu.tw

strong interaction between the social enterprises and the Indigenous communities (Sengupta et al. 2015).

I employ a qualitative approach in this chapter. An in-depth interview with the leader of a Cordillera-based social enterprise serves as my primary data source. In addition, a literature review provided a theoretical framework and contextual understanding of social entrepreneurship, cultural preservation, historical education, and socioeconomic development. It draws upon scholarly articles, academic books, reports, and relevant publications to establish a strong foundation for the study's analysis and discussion. The following questions will be discussed in this paper: In what ways does the Philippine government support returning OFWs? Can social enterprises provide a reliable livelihood for indigenous migrant workers who want to return to their home country? I utilize discourse analysis to answer these questions. To contextualize this research, I briefly review the literature on return migration. To further assess how much this entrepreneurial model may contribute to the long-term sustainability of the Indigenous Peoples, I examine the literature on social entrepreneurship. Then, taking Cordillera as a case study, I analyze the effectiveness of policies designed to spur economic expansion and the viability of social enterprises' efforts to reintegrate into local communities. I argue that working for social enterprises can benefit returning Indigenous OFWs because it fosters a sense of community belonging within a social enterprise setting and provides economic and social empowerment that is intrinsically linked to their identity, cultural heritage, and design. Although much has been said about the importance of social entrepreneurship, I seek to contribute to the discussion by looking at how social entrepreneurship can be a viable livelihood program for returning OFWs, with a particular emphasis on the Indigenous migrant workers.

## 2 Indigenous Migrant Workers

Indigenous migrant workers represent a distinct subset within the broader context of Overseas Filipino Workers (OFWs). The estimated number of Overseas Filipino Workers (OFWs) or Filipinos working abroad between April and September 2022 was 1.96 million. Among the country's 17 regions, the Cordillera region accounted for 1.8% of these OFWs (PSA 2023). According to the 2020, the indigenous population identified by the National Commission on Indigenous Peoples (NCIP) comprised 9.84 million or 9.1% of the 108.67 million household population. The Philippines is home to various ethnic groups, each with its own unique cultural heritage. One such group is the Cordillerans, who live in the mountainous Cordillera region in the northern part of the country.

Traditionally known as the Igorots, the indigenous communities from the provinces of Kalinga Apayao, Mountain, Ifugao, Abra, and Benguet prefer to identify as "Cordillerans," a term that was considered subversive by the state until recent times (Hyndman 1991). In this chapter, I will use the term Cordillerans to refer to these indigenous peoples. The Cordillerans possess a diverse array of traditions and customs that have been passed down through generations. Their culture is closely connected to the natural environment they inhabit, reflecting a deep respect for nature. Through their festivals, craftsmanship, rituals, and oral traditions, they contribute significantly to the cultural richness of the Philippines, offering valuable insights into the nation's diverse heritage.

Most Cordillerans have inhabited these mountainous regions since ancient times, largely isolated from the rest of Filipinos. This isolation has facilitated the preservation of their distinct cultures despite centuries of colonization by various external forces. However, indigenous peoples are persistently regarded as the most disadvantaged members of society, facing significant socioeconomic challenges due to limited literacy levels, which result in elevated rates of underemployment and unemployment.

The scarcity of job opportunities or the inadequacy of wages in the Cordillera region has prompted both indigenous and non-indigenous Filipinos to seek employment opportunities abroad. The allure of higher salaries and better living conditions motivates individuals to embark on overseas migration to improve their economic prospects and support their families back home. In the face of limited options within the local job market, migration becomes not only a pragmatic choice but often a necessity for many households.

Families rely on the remittances sent by migrant relatives to meet basic needs, such as food, education, and healthcare. Consequently, migration has become deeply entrenched in the social and economic fabric of indigenous communities in the Cordillera region, mirroring broader migration trends observed across the Philippines. However, the decision to migrate is not without its challenges and risks.

Indigenous migrant workers face numerous obstacles, including navigating complex migration processes, adapting to unfamiliar cultures and environments, and coping with separation from their families and communities. Moreover, the nature of low-skilled work abroad often exposes migrant workers to exploitation, abuse, and precarious living conditions. Despite these challenges, the prospect of earning higher incomes and providing a better future for their families compels many Indigenous individuals from the Cordillera region to pursue opportunities overseas.

## 3 Return Integration

Since labor migration is always perceived as temporary, a return journey is regarded as a logical next step in the process (Asis 2008). Cerase (1974) and Cassarino (2014), on the other hand, accurately underline conditions that impact migrant workers' expectations before the departure, as well as their psychological and financial preparation upon return to their home country. According to Kuschminder (2017), despite the rising demand for migrants to return and reintegrate into their "home countries," the mechanism is still limited, and scholarly emphasis is increasingly shifting to address this gap. Even though these studies have centered our attention on return migration, there is still a need to study the reasons for it and learn about the pressure to reintegrate migrant workers, which is still rigorously imposed on the degree of preparation of the migrants' returns in their home country (Nisrane et al. 2017; Scalettaris and Gubert 2018). Researching the pressures and difficulties experienced by migrant workers after return is also significant. Many migrants who return to their home countries face challenges reintegrating into society, including difficulties securing employment. These difficulties may be exacerbated if the return was unanticipated or if the migrant's foreign-acquired skills and credentials are not accepted or applicable in their home country. By recognizing and addressing these challenges, governments and organizations may better provide returning migrants with the resources they need to successfully reintegrate into society and improve their quality of life successfully.

Some studies suggest that returning migrants can boost their home countries' economies (Cai 2011). Studies have indicated that many returnees would rather start their own businesses than return to the workforce (Constant and Zimmermann 2004). According to McCormick and Wahba (2001), differences in work prospects abroad considerably impact the likelihood of returning Egyptian migrants being entrepreneurs when they return back home. Foster (1967) examines how the increased wealth amassed by Mexican Braceros working in the United States impacted the level of life in their home communities upon their return to Mexico. International migration and its effects on the economic and social development of six rural Egyptian communities are examined by Reichert (1993). In the study, the researcher found that the more rural a village was, the more likely it was to have households with at least one migrant member. However, return migrants may benefit local economies by investing in assets and starting businesses with the money they saved and the knowledge they gained during their time away (Démurger et al. 2008; Démurger and Xu 2011; Yu et al. 2016). The money that migrants save while working abroad is a major benefit for those who decide to return home. Migrants often save a significant portion of their earnings, intending to invest in their home countries upon their return. The findings of Démurger et al. (2008) reveal that rural migrants often work in low-end, non-tenured public sector positions that pay extremely little on average but do provide some returns to schooling. Using a unique rural household survey from Wuwei County, Démurger and Xu (2011) examine the factors that influence returning migrants' decisions to get into self-employment upon their return to their hometowns. Their research suggests that returning to one's hometown might assist in stimulating rural economies and reduce poverty in China's less developed regions. In their article, Yu et al. (2016) investigate the link between leaving the countryside for the city and starting a business. Their research shows that migrants improve their prospects of starting their own businesses by gaining access to a broader network of contacts and a larger pool of talent and resources. This data supports the hypothesis that resettling migrants' efforts to start businesses in rural areas might help stimulate economic growth and reduce poverty.

According to the other available research, returning to one's country of origin does not lead to any notable growth. Some studies have shown that even after working in another country for many years and earning far more money than they could back home, migrant workers often end up back in their native country completely destitute. According to Saguin (2020), transnationalism explains the unwillingness of migrant workers to save. His research shows that migrants spend their money on reintegrating into society rather than saving for the future. Instead of putting money down for their own retirement, they utilize their earnings to prepare their family for their eventual return home via means such as remittances and frequent visits. After several years of working overseas, one in ten migrant workers go back to their home countries, which are still financially unstable, according to research by Basa et al. (2012). Several people experienced problems adjusting to their previous social circles, including their families, and obtaining stable employment (ADB 2013). Entrepreneurs who have returned from migrating face numerous obstacles and challenges. While there is encouragement for migrant workers to embark on business ventures, many encounter failure. The repercussions of worldwide crises further exacerbate the plight of returning migrant workers.

## 4   Reintegration in the Philippines

Supporting returning migrants is a policy priority for the Philippine government because it recognizes Filipino migrant workers as economic agents and because it expects them to be economically engaged as part of their reintegration into the nation (Saguin 2020). To help migrant workers return to the workforce and use their talents to create a stronger country, the legislation mandated reintegration services and established a Replacement and Monitoring Center under the Department of Labor. Under the Republic Act 10022 and its Implementing Rules and Regulations, the Philippines set up the National Reintegration Center for OFWs (NRCO) at the Department of Labor and Employment (DOLE). The Reintegration Services are programs and initiatives designed to help returning overseas workers find employment in their home communities and use their newfound resources to start or expand businesses and provide for their families (Wickramasekara 2019).

The services include various components to support returning overseas Filipino workers (OFWs). Firstly, there is the "psycho-social" component, which includes psycho-social therapy, stress debriefing, values development, and financial literacy, among other services. Secondly, the means of making a living is facilitated through several programs. The Balik-Pinas, Balik-Hanapbuhay Program offers non-cash livelihood support to returning OFWs, both active and inactive, who have been displaced due to war, political conflicts, policy changes, or have been victims of illegal recruitment and human trafficking. The Balik-Pinay, Balik-Hanapbuhay Program also provides women who have returned from working abroad with the necessary tools to start their own businesses, prioritizing household service workers in distress due to war, violence, or trafficking.

Financial Awareness Seminars (FAS) and Small Business Management Training (SBMT) are designed to educate OFWs and their families on managing finances and inspire them to start their own businesses. The Livelihood Development Assistance Program (LDAP) offers grants for livelihood assistance to undocumented returning

OFWs through livelihood starter kits. Similarly, the Education and Livelihood Assistance Program (ELAP) grants undocumented returning OFWs educational and livelihood grants. These programs include scholarships for children of OFWs and financial assistance to surviving spouses or parents of deceased OFWs. According to the Department of Labor and Employment (DOLE) website, these scholarships and financial aid vary in amount, with elementary students receiving P5,000, high school students P8,000, college students P10,000, and surviving spouses or parents receiving P15,000.

Current reintegration initiatives for overseas Filipino workers (OFWs) aim to promote "brain circulation" by creating a setting and infrastructure that would allow OFWs to put their gained skills and knowledge to use upon their return home (Calzado 2007). There were 9902 beneficiaries of livelihood aid for OFWs between 2011 and 2018 (Banta 2020). When migrant workers return to the Philippines, one goal of reintegration is to help them find employment or start their own businesses so that they may once again contribute to the local economy. To achieve sustainable reintegration, Willoughby (2009) argues that a nation should fully harness the potential for growth of the returning migrant at the Global Forum on Migration and Development (GFMD). If migrants are provided with the necessary cultural, social, and economic resources, they will be able to contribute to the growth of their native country (Rashad 2014). Banta (2020) argues that neoliberal development, with its jubilant celebrations of entrepreneurship and harsh elimination of social assistance for its former laborers, is the proper environment to see reintegration as a socio-spatial logic for controlling and regulating the flow of OFWs.

The primary aim of these reintegration initiatives is to facilitate "brain circulation," enabling Overseas Filipino Workers (OFWs) to apply the skills and knowledge acquired abroad upon their return to the Philippines (Calzado 2007). However, despite the inclusive nature of these programs, it is crucial to acknowledge the unique challenges encountered by indigenous migrant workers. Unlike the broader returning migrant population, Indigenous migrants often confront distinct hurdles related to cultural preservation, land rights, and social integration upon their reentry into their home communities.

Hence, it is imperative to establish targeted mechanisms tailored specifically to the needs of indigenous migrant workers. By acknowledging their distinct circumstances and providing specialized support, the Philippine government can ensure their successful reintegration while safeguarding their cultural heritage and fostering socio-economic progress within indigenous communities. These mechanisms should include comprehensive assistance services addressing economic stability and broader requirements such as healthcare, education, and psychological well-being. Effective collaboration between government agencies and civil society organizations is essential for tackling these individual, community, and systemic challenges.

Furthermore, the rise of social enterprises presents an opportunity to promote inclusivity and economic empowerment among marginalized groups, including returning Indigenous migrant workers. The Philippines can bridge the gap between economic growth and poverty reduction, particularly in underserved regions, by creating a conducive environment for social enterprise development. Therefore, implementing specific mechanisms tailored to the needs of returning Indigenous migrant workers is crucial for ensuring their successful reintegration into their community.

## 5   Social Entrepreneurship

Although more and more studies are devoted to the topic, it is difficult to pin down precisely what defines social entrepreneurship. Social entrepreneurship, the solidarity economy, and community economies are all frameworks through which social innovation has been studied in the literature (Moulaert et al. 2014). Concentrating on developing social enterprises will bring about a fresh perspective on the promotion of decent work and the addition of new employment opportunities, all while reducing poverty, enhancing local communities, and giving voice to disadvantaged minorities. Taking this strategy may help reduce obstacles to economic stability, especially for marginalized groups, and it can do so in a manner that is both sustainable and empowering. According to Dacanay (2012), the "paradox of excessive poverty and inequality" in the Philippines may be resolved by shifting from a corporate-led market to an ethical and social economy via social entrepreneurship.

The British Council (2015) undertook a qualitative review of social entrepreneurship in the Philippines. According to the research, social entrepreneurship focuses on the nation's capital and targets low-income people. In the Philippines, social businesses were generally producer-based associations that distributed sales earnings to low-income stakeholders. She found that many community members depend on non-capitalist economic activity. The study of Bebbington (2000) among the indigenous Quichua of the Andes shows how development affects and transforms their local ecosystems. Since "alternatives to contemporary capitalism are being established," he argues, neither post-development nor neo-liberal theories offer a complete depiction of the developments in the region. Using "comparative ethnographies," he analyzes Indigenous cultures' attempts to "build a meaningful existence." His study shows indigenous Quichua tribes can self-govern while adjusting to a globalized economy. Rural residents assume that urbanites have greater work opportunities. Contrary to popular opinion, Bebbington finds that emigration helps individuals invest in their houses and communities in Colta. The residents kept farming and formed neighborhood associations that advocated good governance. The research of these communities shows that market-based development is not always detrimental, and accumulation is not always in conflict with indigenous values. The development process has also boosted community self-government.

The potential for business to 'do good' is far from certain. Several organizations mix 'commercial motivation' and social motives. Social businesses are new and imaginative ways to solve social concerns. Social businesses possess a notable advantage in utilizing market mechanisms and entrepreneurial strategies to address social challenges effectively. These enterprises construct sustainable models that benefit communities by integrating business strategies and inventive solutions, coupled with a comprehensive comprehension of social and cultural contexts. Notably, the involvement of local stakeholders, including indigenous groups, in the decision-making process safeguards cultural and social characteristics, concurrently addressing economic requirements. Around the world, indigenous groups have unique ways of living since "business" and economic activities are ingrained in cultural and social traits (Cahn 2008). These practices often exemplify a harmonious coexistence with nature, a profound sense of community, and a steadfast commitment to sustainability. These enterprises actively embrace their cultural heritage and traditional wisdom, employing them to create distinctive goods and services catering to local and global markets. Such businesses hold immense potential for illuminating alternative economic models that prioritize the well-being of communities, the responsible stewardship of the environment, and the sustainability of future generations.

## 5.1  Indigenous Social Enterprises in Cordillera

Traditional Filipino culture and the contemporary upswing in Filipino handicrafts testify to the Philippines' cultural richness and diversity. Weaving, a traditional textile manufacturing method, has quickly become one of the most visible and successful rural businesses. Weaving on a handloom in the Philippines has deep spiritual roots; it was traditionally believed to be a channel between humans and the afterlife, where it might be used to ensure the welfare of the living and ward against evil. They also make good presents for older people because of their association with luck and protection when passed down through the generations (Labrador 2016). In rural areas, exceptionally skilled artisans have kept weaving alive and well by maintaining and improving upon centuries-old traditions and practices. Textiles these artists make are detailed and visually appealing because of the use of raw materials, including indigenous fibers and plant-based dyes. The finished products are excellent examples of how cultured history may be combined with modern aesthetics by fusing traditional craftsmanship and contemporary design, blending cultural heritage with modern aesthetics.

Since the early 1990s, when the Philippine government under President Fidel V. Ramos pushed initiatives to "fast track rural development," woven textiles from the Cordillera region have developed their distinctive multi-platform nature (Bankoff and Weekley 2002). The primary beneficiaries of business development initiatives are now weavers rather than merchants. Weavers' circumstances have not always improved due to past actions targeting small firms that sell weavers' goods, with some claiming that merchants "exploited" weavers (Suratman 1991). In 1992, key legislation was passed to protect specific cultural traditions. The National Culture Commission on the Arts (NCCA) was founded by Republic Act No. 7356 to encourage communities to record and renew their cultural traditions. Filipinos practicing traditional art are protected by the Manlilikha ng Bayan Act (Republic Act No. 7355).

Subsequently, in 1997, the Indigenous Peoples' Rights Act (Republic Act No. 8371) was passed. The NCCA's flagship initiative, the School of Living Traditions (SLT), aims to maintain cultural traditions by passing on indigenous knowledge and skills from cultural masters to youngsters in local communities. As a result, communities were encouraged to record and reinvigorate their unique cultural traditions. Businesses that can meet the rising international demand for locally produced handmade items have been encouraged, for example, to be founded by individual entrepreneurs and the heads of non-governmental organizations (NGOs). Scholars have paid little attention to the expanding number of social companies in Cordillera working to improve corporate transparency, product quality, producer-buyer relationships, and community well-being. The proliferation of weaving enterprises in Cordillera has resulted in several societal and economic benefits. First, it has given many people a chance to make a living, especially women who have traditionally had essential roles in weaving cultures. By engaging in weaving as a source of income, these artisans contribute to their households' economic stability and the overall prosperity of their communities. Moreover, weaving enterprises often promote inclusive practices by incorporating marginalized groups and empowering them economically. Milgram (2020) examines the phenomenon of social businesses in the Philippine provinces of Ifugao and Benguet. She argues that craft businesses show how social entrepreneurship can be a powerful force for positive change in a community and its people's standard of living. Because of the efforts of a set of social entrepreneurs, the living and working conditions of Cordillera artisans have improved, and a new collection of uniquely designed textiles has been launched. Some businesspeople with a social conscience in the Cordillera have raised standards of living for artisans and created a new line of beautifully crafted textiles.

The primary economic activity in the Cordillera provinces, similar to the other rural regions across the Philippines, revolves around cultivating wet rice and vegetables. Beyond weaving, Cordillera hosts diverse social enterprises, including agritourism ventures, ecotourism initiatives, handicraft production, and community-based conservation projects, all of which contribute to the region's economic and social development.

## 5.2   Weaving in Cordillera Region

If there is one distinctive aspect that the Cordillera region is renowned for, it would be their elaborate handwoven patterned textiles. The weaving industry in the Cordillera region of the Philippines is a vibrant and culturally significant sector that embodies indigenous communities' rich heritage and traditions. This industry not only provides a source of livelihood for many artisans but also plays a crucial role in preserving the unique cultural identity of the Cordillera people. Salvador-Amores et al. (2021) argue that Cordillera weaving practices have developed into a platform for cultural expression and economic activities for local communities. More and more Cordillerans are engaging in the weaving industry.

The people in the Cordillera are receiving government support to revitalize and promote the weaving industry in the region. In 2022, the Regional Development Council endorsed the establishment of the Cordillera Weaving Industry Council (CWIC) to propel the sector's growth. The council aims to sustain assistance for local weavers and bolster industry promotion, acknowledging its evolution into an economic enterprise rather than solely a cultural asset. Building on this momentum, in 2021, the council adopted the Cordillera Weaving Industry Development Plan (CWIDP) 2020–2028 as a regional policy blueprint. Crafted collaboratively by various stakeholders, including indigenous weavers, entrepreneurs, regional agencies, fashion designers, and RDC Committees, the CWIDP seeks to cultivate a dynamic and thriving weaving sector. It prioritizes enhancing weavers' skills in production technology innovation and strengthening market connections. Furthermore, the plan advocates for environmental quality, sustainable resource use, resilience-building for individuals and communities, and preserving the region's unique culture and heritage.

2021 represents a significant milestone in the region's efforts to support the weaving industry comprehensively. This strategic framework was developed collaboratively by various stakeholders, including indigenous weavers, weaving entrepreneurs, regional line agencies, Cordillera fashion designers, and the RDC Committees on Economic Development and Indigenous Peoples Concerns. The CWIDP is designed to foster a dynamic and thriving weaving sector by addressing critical areas of development.

The CWIDP strongly emphasizes enhancing the skills of weavers through investment in production technology innovation and establishing more robust connections to markets. By doing so, it aims to improve the quality and efficiency of weaving practices, expand market reach, and increase economic opportunities for local weavers. Additionally, the plan advocates for environmental sustainability and the responsible utilization of natural resources, recognizing the importance of preserving the region's ecological integrity for future generations.

## 5.3   Reintegration of Indigenous Migrants Through Social Enterprises

Indigenous Peoples are disproportionately more likely to live in poverty than the rest of the Filipino population. In the 1980s, sending workers overseas became legally institutionalized, giving indigenous peoples another option for finding employment outside of their own countries. The study of Banta (2020) includes Indigenous Bugias people who took a chance and worked as domestic servants in locations like Israel, Cyprus, and Hong Kong, sending money back to their families in the mountains so that they may use it to construct permanent structures there. The substantial contributions made in investigating the connections between international migration and agricultural development are mapped out in several studies. How female migrant laborers and their remittances have changed rice fields in the Ifugao province of the Cordillera is the subject of research by McKay (2003). She contends that migration facilitates the construction of "new economic identities and new views of the future" that are motivated by local economic prospects and individual aspirations about accumulating cultural capital and internalized ideologies of modernization and progress.

Establishing and expanding new enterprises in rural regions can boost regional economies by generating new employment opportunities, increasing access to a broader range of goods and services, and enhancing residents' overall quality of life (Yu et al. 2016). A local bias in entrepreneurship is a hypothesis proposed by Michelacci and Silva (2007). Individuals who are born in a particular region have a competitive edge over non-native workers when initiating new businesses due to their more profound understanding of the local financial opportunities and social connections. Native residents, in contrast to migrants, are more inclined to create inventive business approaches that align with the strengths of their communities. This phenomenon can be attributed to the transfer of knowledge and technology, which significantly contributes to fostering entrepreneurship among the indigenous population. The exploration of indigenous peoples' entrepreneurial activities has been examined as a possible driver for societal and economic transformation. Numerous books and articles explore the achievements of Indigenous entrepreneurs. Colbourne (2017) uses a case study technique to analyze entrepreneurship at Manitobah Mukluks, a Canadian company that develops and manufactures mukluks and moccasins and is frequently lauded for its innovations. Since Sean McCormick's foundation, the company has focused on helping Indigenous artists and communities in Winnipeg, Manitoba, and throughout Canada (Pauls 2015). The success of Manitobah Mukluks exemplifies how entrepreneurs can draw on their unique perspectives, experiences, and identities to develop innovative products and services that benefit their communities by perpetuating Indigenous values, culture, and traditions (Smith 2015). Success rates for entrepreneurship and hybrid enterprises increase when the rights of Indigenous peoples are recognized and protected and when Indigenous communities either take the lead or are actively involved in the efforts.

The investigation of Cahn (2008) on the topic of entrepreneurship highlights another effective model. According to her research on the connections between micro-enterprises and *fa'asamoa* in rural communities of Samoa, an "indigenous" style of company had evolved where the two melded effectively, and the success and sustainability of the micro-enterprise were enhanced. These companies take advantage of the growing international recognition of Indigenous peoples' rights and autonomy to help Indigenous communities pursue economic decolonization and economic reconciliation through the refocusing of development efforts on Indigenous centered on social, cultural, economic, and environmental value creation (Sengupta et al. 2015).

Entrepreneurship offers significant benefits for returning Indigenous migrant workers by focusing on economic empowerment, cultural preservation, and social integration. They generate jobs suited to local expertise, support business initiatives, and highlight traditional practices, helping maintain cultural heritage. These enterprises provide comprehensive support, easing social reintegration and reducing feelings of isolation. Their commitment to sustainable practices ensures environmental care and reinvestment in community development.

Returning Indigenous migrant workers can join social enterprises as employees or set up their own if they have the capital. They can engage in various sectors that leverage their skills, experiences, and cultural heritage. Agriculture and agribusiness offer significant opportunities through organic farming, agroforestry, and value-added agriculture, enabling workers to apply traditional farming techniques and local crop knowledge. The tourism and ecotourism sectors provide avenues for cultural tourism, offering authentic cultural experiences such as traditional crafts, music, and dance, as well as ecotourism, which promotes environmental conservation and respect for local cultures. Additionally, the handicrafts sector allows workers to utilize traditional skills in weaving, pottery, and other crafts, creating unique products for local and broader markets. Homestay tourism is also slowly thriving in Cordillera. Another opportunity for returning migrants is homestay tourism in Cordillera, which the Philippine government actively promotes. Homeowners are encouraged to convert their houses into homestay establishments, providing a valuable alternative when traditional accommodation options are scarce, especially during peak tourist seasons. Homestays offer visitors an authentic experience, immersing them in local culture and lifestyle. This initiative addresses the accommodation shortage, generates additional income for local families, and promotes community-based tourism. By leveraging existing homes, the government aims to enhance the province's tourism infrastructure, ensuring visitors have comfortable and culturally enriching places to stay.

## 6  Case Study: Weaving Entrepreneurship in Cordillera

I have interviewed a representative from one of the well-known social enterprises in Cordillera. During the interview with the head of the social enterprise in Cordillera, it became evident that their organization embodies a multifaceted approach beyond mere profit-making. The interviewee emphasized that their enterprise is driven by a broader mission encompassing several interconnected goals. Firstly, the enterprise actively promotes the preservation of cultural heritage within the Cordillera region. They play a pivotal role in ensuring the continuity of these invaluable cultural traditions by supporting traditional practices, such as weaving, pottery, or traditional arts and crafts. Through their initiatives, the enterprise seeks to safeguard the unique cultural identity of the Cordillera communities and prevent the erosion of traditional knowledge and skills.

He explains that establishing social enterprises revitalizes traditional weaving. In the process, there was a serious realization that for heritage conservation to work, the economic aspect of artisans needs to be addressed. Improving the earning capacity of weavers results in the revitalization of traditional weaving. To date, there are 65 members of their attached organization, which is a member organization behind their social enterprises.

Moreover, the social enterprise is committed to education about the history and significance of the Cordillera culture. The interviewee stressed the importance of raising awareness and fostering a deeper understanding of the region's rich cultural heritage among the local population and a wider audience. He mentioned that workshops, exhibitions, and educational programs aim to shed light on Cordilleran indigenous communities' historical struggles, achievements, and contributions. By providing platforms for knowledge dissemination and historical consciousness, the enterprise empowers individuals to connect with their roots, instills a sense of pride, and strengthens the community's cultural fabric. He said that, in a broader aspect, the income derived by the organization is used to maintain its education center, a community institution on the premises of the school in its area. This center delivers heritage education to teachers, students, and community members. It currently houses a weaving center, two galleries on Ifugao material culture, and a small Ifugao library. The center provides training for teachers on integrating Ifugao indigenous knowledge systems and practices into the formal education system. The center is also a venue for cultural community events, practices that are slowly disappearing due to the onslaught of modernity. The center envisions being a prime mover in conserving the tangible and intangible culture of the Ifugaos.

Social entrepreneurship has been studied as a means of long-term economic growth for Indigenous communities. Studies by Dana and Light (2011) analyze the role of social enterprises in providing economic and social opportunities for Indigenous people. In addition, having a foundation in the local community helps Indigenous people protect and grow their social and cultural assets (Ratten and Dana 2015). Boosting the earning capacity of weavers has been a central objective of social enterprises in the Cordillera region, exemplified by the case of the social enterprises I interviewed. As a social enterprise, it is unique because the weavers' organization owns it. This ownership structure ensures that the weavers not only receive compensation for their individual labor but also benefit from being members of the organization. Since establishing their brand, member-weavers have experienced a significant increase in their earnings. They are now paid three times the amount they used to receive from middlemen or souvenir shop owners. This substantial increase in compensation serves as an incentive for the weavers to actively participate in the organization's initiatives and commit to producing high-quality traditional textiles.

One distinguishing feature of this social enterprise is its commitment to prompt payment to weavers. Upon submitting their finished products, weavers are assured of immediate payment, a departure from the traditional consignment model, where weavers often have to wait until items are sold to receive payment. This practice supports the economic well-being of individual weavers and contributes to the overall financial stability of their families and communities. By guaranteeing timely payment, their social enterprise empowers weavers by providing them with a stable and reliable source of income. This practice supports the economic well-being of individual weavers and contributes to the overall financial stability of their families and communities. The increased earnings enable the weavers to invest in their craft, purchase necessary materials, and improve their living standards.

Furthermore, the shift from reliance on middlemen to ownership and direct engagement with the market has additional benefits. This social enterprise bypasses the exploitative practices often associated with the traditional supply chain by eliminating middlemen. Weavers have more control over the pricing and sale of their products, ensuring that their craftsmanship is appropriately valued and recognized. This shift also fosters a sense of pride and ownership among the weavers, reinforcing their commitment to preserving their cultural heritage and promoting the Ifugao textile industry. Under the consignment agreement, the weaver only gets paid when an end user buys the item, which usually takes weeks or months. To illustrate, the organization pays P1500–4000 for every warp of cloth woven on the backstrap. In the Banaue area, that would only cost between P200 and P250. The organization also provides for the materials used by the weaver, again in contrast with the old system where the weaver had to provide her own yarns. Furthermore, finishing a single social enterprise product is not a one-person work. At least three people are involved in the entire process, which may include a weaver, an embroiderer, a beader, and a seamstress who does the final touches. For an item that costs approximately P4000.00, the typical breakdown of actual labor and material costs is as follows: The weaver's fee is P2500, the embroiderer's fee is P500, and the cost of materials such as cotton and dye amounts to P500. Additionally, the seamstress is paid P200. Finally, a markup of P300 represents the organization's net gain per item. He elaborates that their social enterprise "attempts to develop a truly social company by emphasizing the benefit (labor cost) that goes to craftsmen above corporate profit for the organization." Additionally, the business acknowledged the economic difficulties experienced by the Cordillera towns and worked to ameliorate those difficulties. The interviewee described how their organization provides financial relief by creating income-generating opportunities for individuals within the community. They emphasized fair trade practices, cooperative initiatives, and capacity-building programs that enable local artisans and producers to earn a sustainable income from their traditional crafts or products. By empowering individuals economically, the enterprise contributes to poverty reduction, enhances economic resilience, and improves the overall livelihoods of the Cordillera communities.

Giovannini and Davidson-Hunt (2012) explore the nature of Indigenous social enterprises, which they define as "community-based economic activity with a social purpose," and find that such businesses typically maintain the participation of Indigenous people while also resolving particular economic and social issues facing the community. Entrepreneurship is emphasized by Dana and Dana (2005) as an essential aspect of communal life that may foster empowerment via economic activity. Helping Indigenous people become economically independent is critical if they are to achieve political autonomy. The desire to preserve natural resources, strengthen local communities, and sustain long-standing ways of life are hallmarks of many Indigenous business ventures (Dana and Light 2011). Social entrepreneurship emerges as a potential driver of significant community and livelihood development, as illustrated by the craft companies I explore here. The interviewee emphasizes how the Ifugao people's cultural practices, such as Ifugao terrace farming and weaving, are grounded in need and practicality. It also requires the concerted efforts of the local community to preserve it and provide it with the cultural significance it deserves as a component of its legacy. Culture is maintained via communal effort, whereas a social enterprise is distinguished by its commitment to some kind of social good. So, social enterprises may help preserve and advance culture in the future. In addition, he disclosed that the social enterprise now employs five returned overseas Filipino workers (OFWs) in the weaving industry. In addition to his work with Indigenous social entrepreneurs, the interviewee is exploring the potential benefits of social firms for returning OFWs. As an employee of a social company, an OFW is more likely to experience the warm embrace of family and the comfort of community. These social and psychological needs are significant for long-term expatriates. The business might be a way for those who have been away for a long time to find their way back home and reconnect with family, friends, and their culture.

Collaboration between economic and cultural entities is essential to the growth and development of the sector. Government policies that prioritize Indigenous entrepreneurship are crucial. The Department of Trade and Industry (DTI) annually sponsors several trade shows that offer excellent promotional opportunities (Manila FAME, CITEM, et cetera.). However, this is not enough to support the weaving industry as a whole, much less individual weavers or smaller cooperatives. Since the DTI's current advertising approach is not helping the weavers, the interviewee has been recommending that they try something new. When they help privately owned SMEs, they are just helping the rich. Most weavers would not be paid fairly for their work, and they typically would not benefit much from the distribution system they helped create.

Beyond their monetary benefits, social enterprises have vast untapped potential. They serve as platforms for preserving and revitalizing traditional knowledge and cultural practices. Artisans and communities perpetuate their cultural heritage by actively engaging in weaving, fostering a sense of pride and identity. Safeguarding cultural practices in this way protects intangible cultural heritage, strengthens ties to family history, and fosters the passing down of knowledge from one generation to the next.

# 7 Conclusion

In this chapter, I explored the dynamics of social entrepreneurship, Indigenous entrepreneurship, and return migration entrepreneurship. Social enterprises favor the inclusion of the most vulnerable and marginalized people and provide them with opportunities for economic participation. Scholars agree that this can give those people a sense of community and economic and social empowerment deeply tied to identity, cultural heritage, and design.

By situating place-based production and trade within broader discourses of meaning and agency, social entrepreneurs and craftspeople in the textile sector of the Cordilleras may build a sustainable livelihood that is more attentive to their shared immediate needs. In the Philippines, like everywhere in the world, indigenous communities are marginalized. Indigenous peoples get a head start in the industries where the area has a natural competitive advantage because of their wealth of cultural knowledge and natural resources. The academic community agrees that social entrepreneurship may help the Indigenous people create a more stable economic future. I illustrated that sustainable livelihood companies formed via social entrepreneurship have great potential and should be supported by government agencies as part of their reintegration strategies for Indigenous migrant workers.

In conclusion, social enterprises play a crucial role in enhancing the earning capacity of weavers in the Cordillera region. Through ownership and direct engagement with the market, these enterprises empower weavers by providing fair compensation and prompt payment for their labor. This shift away from traditional consignment practices and reliance on middlemen boosts the economic well-being of weavers and fosters a sense of pride, cultural preservation, and sustainable development in the community.

# References

ADB (2013) Impact of global crisis on migrant workers and families: gender perspective. Retrieved from https://www.adb.org/publications/impact-global-crisis-migrant-workers-and-families-gender-perspective

Arceo N (2020) OFW remittances hit record high in December. The Manila Times. Retrieved from https://www.manilatimes.net/2023/02/16/news/national/ofw-remittances-hit-record-high-in-december/1878804

Asis M (2008) The Philippines. Asian Pac Migr J 17(3):349–378. https://doi.org/10.1177/011719 680801700308

Bankoff G, Weekley K (2002) Post-colonial national identity in the Philippines: celebrating the centennial of independence. Ashgate, Aldershot

Banta V (2020) Risky investments: turning return migrants and national heroes into entrepreneurs in the Philippines (Doctor of Philosophy—Ph.D.). University of British Columbia

Basa C, De Guzman V, Marchetti S (2012) International migration and over-indebtedness: the case of Filipino workers in Italy. International Institute for Environment and Development

Bebbington A (2000) Reencountering development: livelihood transitions and place transformations in the Andes. Ann Assoc Am Geogr 90(3):495–520. Retrieved from http://www.jstor.org/stable/1515526

British Council (2015) A review of social enterprise activity in the Philippines. Retrieved from https://www.britishcouncil.org/sites/default/files/social_enterprise_activity_philippines.pdf.

Cahn M (2008) Indigenous entrepreneurship, culture and micro-enterprise in the Pacific Islands: case studies from Samoa. Entrep Reg Dev 20(1):1–18. https://doi.org/10.1080/089856207015 52413

Cai F (2011) The labour export policy: a case study of the Philippines. Retrieved from https://www.e-ir.info/2011/08/24/the-labour-export-policy-a-case-study-of-the-philippines-2/#google_vignette

Calzado R (2007) Labour migration and development goals: the Philippine experience. Paper presented at the International Dialogue on Migration, WMO Conference Center, Geneva

Cassarino J-P (2014) A case for return preparedness. In: Global and Asian perspectives on international migration, pp 153–165

Cerase FP (1974) Expectations and reality: a case study of return migration from the United States to Southern Italy. Int Migr Rev 8(2):245–262. Retrieved from https://www.jstor.org/stable/300 2783

Colbourne R (2017) Indigenous entrepreneurship and hybrid ventures. In: Hybrid ventures, pp 93–149

Constant AF, Zimmermann KF (2004) The making of entrepreneurs in Germany: are native men and immigrants alike? Retrieved from Bonn

Dacanay ML (2012) Social enterprises and the poor: enhancing social entrepreneurship and stakeholder theory (Doctoral School of Organisation and Management Studies). Copenhagen Business School. Retrieved from https://research-api.cbs.dk/ws/portalfiles/portal/58899306/Marie_Lisa_Dacanay.pdf

Dana LP, Light I (2011) Two forms of community entrepreneurship in Finland: are there differences between Finnish and Sámi reindeer husbandry entrepreneurs? Entrep Reg Dev 23(5–6):331–352. https://doi.org/10.1080/08985626.2011.580163

Dana LP, Dana T (2005) Expanding the scope of methodologies used in entrepreneurship research. Int J Entrepreneurship Small Bus 2(1):79–88. Retrieved from https://doi.org/10.1504/IJESB.2005.006071

Démurger S, Gurgand M, Li S, Ximing Y (2008) Migrants as second-class workers in urban China? A decomposition analysis. J Comp Econ 37:610–628. https://doi.org/10.1016/j.jce.2009.04.008

Démurger S, Xu H (2011) Return migrants: the rise of new entrepreneurs in rural China. World Dev 39(10):1847–1861. https://doi.org/10.1016/j.worlddev.2011.04.027

Foster G (1967) Tzintzuntzan: Mexican peasants in a changing world. Little Brown & Company, Boston

Giovannini M, Davidson-Hunt IJ (2012) Social enterprises for development as buen vivir. J Enterprising Commun People Places Glob Econ 6(3):284–299. https://doi.org/10.1108/175062012 11258432

Hyndman D (1991) Organic act rejected in the Cordillera: dialectics of a continuing fourth world autonomy movement in the Philippines. Dialect Anthropol 16:169–184. https://doi.org/10.1007/BF00250244

Kuschminder K (2017) Reintegration strategies: conceptualizing how return migrants reintegrate. Palgrave Macmillan, Cham, Switzerland. 978-3-319-55741-0

Labrador AP (2016) Hibla ng Lahing Filipino: the artistry of Philippine textiles, 2nd edn. In: Museum N, Legarda OOSL (eds). Manila, Philippines

McCormick B, Wahba J (2001) Overseas work experience, savings and entrepreneurship amongst return migrants to LDCs. Scott J Polit Econ 48(2):164–178. https://doi.org/10.1111/1467-9485.00192

McKay D (2003) Cultivating new local futures: remittance economies and land-use patterns in Ifugao, Philippines. J Southeast Asian Stud 34(2):285–306. https://doi.org/10.1017/s00224634 03000262

Michelacci C, Silva O (2007) Why so many local entrepreneurs? Rev Econ Stat 89(4):615–633

Milgram BL (2020) Fashioning frontiers in artisanal trade: social entrepreneurship and textile production in the Philippine Cordillera. South East Asia Res 28(4):413–431. https://doi.org/10.1080/0967828x.2020.1834336

Moulaert F, MacCallum D, Mehmood A, Hamdouch A (2014) The international handbook on social innovation: collective action, social learning, and transdisciplinary research. Edward Elgar Publishing Limited, Cheltenham, UK

Nisrane BL, Morissens A, Need A, Torenvlied R (2017) Economic reintegration of Ethiopian women returned from the middle east. Int Migr 55(6):122–136. https://doi.org/10.1111/imig.12358

Pauls K (2015) Manitobah Mukluks: building success from the ground up. CBC News. Retrieved from https://www.cbc.ca/news/canada/manitoba/manitobah-mukluks-building-suc cess-from-the-ground-up-1.3197893

PSA (2023) 2022 survey on overseas Filipinos (final result). Philippine Statistics Authority. Retrieved from https://www.psa.gov.ph/statistics/survey/labor-and-employment/survey-ove rseas-filipinos/node/1684061314

Rashad L (2014) The assistance of the government services, civil society, and social network during the reintegration of returned female domestic workers and entertainers in Manila. Utrecht University, Netherlands

Ratten V, Dana LP (2015) Indigenous food entrepreneurship in Australia: Mark Olive 'Australia's Jamie Oliver' and Indigiearth. Int J Entrep Small Bus 26(3):265–279. https://doi.org/10.1504/ ijesb.2015.072391

Reichert C (1993) Labour migration and rural development in Egypt: a study of return migration in six villages. Sociol Rural 33(1):42–60. https://doi.org/10.1111/j.1467-9523.1993.tb00946.x

Saguin K (2020) Returning broke and broken? Return migration, reintegration and transnational social protection in the Philippines. Migr Dev 9(3):352–368. https://doi.org/10.1080/21632324. 2020.1787100

Salvador-Amores A, Martin M, Acabado S (2021) Expressive cultures: empowering Cordillera (Philippines) weavers through textile revitalization. In: Acabado S, Kuan D-W (eds) Indigenous peoples, heritage and landscape in the Asia Pacific knowledge co-production and empowerment. Routledge

Scalettaris G, Gubert F (2018) Return schemes from European countries: assessing the challenges. Int Migr 57(4):91–104. https://doi.org/10.1111/imig.12467

Sengupta U, Vieta M, McMurtry JJ (2015) Indigenous communities and social enterprise in Canada. Can J Nonprofit Soc Econ Res 6(1):104–123. https://doi.org/10.22230/cjnser.2015v6n1a196

Smith R (2015) For Aboriginal peoples, entrepreneurship is the path to economic independence. Accessed at The Globe and Mail. Retrieved from https://www.theglobeandmail.com/opinion/ for-aboriginals-entrepreneurship-is-the-path-to-economic-independence/article24327664/

Suratman S (1991) "Weaving" a development strategy: cottage industries in the Philippines. J Soc Issues Southeast Asia 6(2):263–289. http://www.jstor.org/stable/41056826

Wickramasekara P (2019) Effective return and reintegration of migrant workers with special focus on ASEAN member states. Retrieved from Bangkok, Thailand https://ssrn.com/abstract=357 9358, https://doi.org/10.2139/ssrn.3579358

Willoughby J (2009) Preparing contract workers for return and reintegration—relevant for development? Global Forum on Migration and Development, Athens

Yu L, Yin X, Zheng X, Li W (2016) Lose to win: entrepreneurship of returned migrants in China. Ann Reg Sci 58(2):341–374. https://doi.org/10.1007/s00168-016-0787-0

# Evolving Partnerships for Development: Transnational Agrarian Movement and Farmer Empowerment in the Philippines

Renee Tila

**Abstract** This qualitative study explores the transformation of transnational agrarian movements, exemplified by the Federation of Independent Organizations in Rural Areas in Mindanao (UNORKA-Mindanao), into "partners for development." It examines the influence of local socio-political factors on UNORKA's farmer empowerment strategies in small-scale banana-growing communities in Davao del Norte from 2000 to 2023. The study aims to assess the extent to which these strategies facilitate farmer empowerment. In-depth interviews with four UNORKA officers, two government officials, and three associate farmer cooperatives reveal that UNORKA's national campaigns for land distribution and land occupation tactics were crucial in enabling small-scale banana farmers to assert their rights to land and livelihoods. However, challenges related to legal claims and productive activities persist due to deeply rooted anti-reform sentiments among traditional authorities and local elites. This prompted UNORKA to redirect its initiatives toward capacity-building and rural development projects to support small farmers' livelihoods. UNORKA's experience underscores the significance of empowering small-scale banana farmers in the Philippines, whose livelihoods are intricately linked to the global agribusiness industry. The findings presented in this chapter contribute to the ongoing discussion on the role of civil society organizations in effective strategies for promoting land rights and social justice in agrarian contexts.

**Keywords** Rights-based agrarian reform · Social justice · Rural development · Grassroots movements

R. Tila (✉)
Graduate School of Economics, Kyoto University, Kyoto, Japan
e-mail: reneetila@gmail.com

© The Author(s), under exclusive license to Springer Nature Singapore Pte Ltd. 2024     177
A. Sharifi et al. (eds.), *Navigating Peace and Sustainability in an Increasingly Complex World*, World Sustainability Series, https://doi.org/10.1007/978-981-97-8772-2_9

# 1   Introduction

Transnational Agrarian Movements (TAMs) have advocated for peasant rights since the late 1980s. La Vía Campesina (LVC) is the leading international network of peasant and small farmer organizations, with more than 182 member organizations across 81 countries. LVC plays a crucial role in empowering peasants to participate in global forums and challenge global institutions to promote their rights and interests, as well as those of small-scale farmers worldwide (Borras et al. 2008; Edelman 2024).

Peasants play a significant role as agents of change and are important partners in advocating for agrarian reform. Their involvement in advocacy campaigns and grassroots movements strengthens the collective effort to secure land rights and bring social change (Borras 2008). By focusing on land-related issues, the movements empower rural communities to challenge oppressive structures that perpetuate land inequality (Huizer 2001). When mobilized, peasants have significant political influence (Rigg 2024), making their alliance with organizations like LVC crucial for political engagement.

Politics consists of the debates, conflicts, decisions, and cooperation among individuals, groups, and organizations regarding the control, allocation, and use of resources, as well as the values and ideas underlying those activities. These values and ideologies shape, legitimize, or challenge how resources are used and distributed (Kerkvliet 2002). Kerkvliet (2013) argues that peasant politics, often rooted in everyday struggles for resources, rights, and recognition, are shaped by diverse socio-political dynamics such as power structures, class relations, cultural norms, and historical legacies. Despite being marginalized and often disenfranchised, peasants navigate these dynamics through various forms of politics, which can be categorized into three main types: official politics, everyday politics, and advocacy politics.

Official politics operate within formal institutional structures and legal frameworks, focusing on governance, legislation, and public administration. Everyday politics consist of informal, subtle forms of political interactions embedded in daily life, which reflect individuals' agency, resistance, and negotiation of power dynamics. Advocacy politics involve organized efforts to influence decision-makers, advocate for specific policies, and address systemic issues through formal political participation (Kerkvliet 2013). Scott (1985) also refers to them as "weapons of the weak." They are distinct from outward forms such as rebellions, are passive, spontaneous, stop short of collective outright defiance, "require little to no coordination or planning," rely on individual actions, and are aimed more at self-help than achieving reforms (Tadem 2009, p. 6). Essentially, peasant politics empowers farmers to organize, advocate, and gain control over their lives and livelihoods.

Government institutions lead official politics, while everyday and advocacy politics involve a broader range of participants, including ordinary individuals, grassroots organizations, and civil society groups. These forms of politics intersect and influence each other as peasant groups navigate both formal and informal channels to address the injustices they face. Borras (2008) describes how LVC used both official and advocacy politics in its campaign for agrarian reform. LVC organized coordinated militant actions at the local, national, and international levels, such as protests, land occupations, and public shaming, to challenge neoliberal land policies and promote agrarian reform. Simultaneously, they engaged in negotiations and partnerships with agencies like the International Fund for Agricultural Development (IFAD) and the Food and Agriculture Organization (FAO) to influence policymaking, expand political spaces for civil society, and develop aligned projects and conferences. This combined approach of direct action and collaboration enables LVC to build alliances, access resources, and effectively advance its agenda for social justice, community control over land, and sustainable agricultural practices. Despite these efforts, the impact of LVC's campaign on policy reform has been described as "marginal." The difficulties encountered by LVC, despite being the largest social movement, in securing commitments from international agencies and official support from state actors demonstrate the substantial impact of local socio-political dynamics on the effectiveness of efforts to influence agrarian reform policies.

Transnational agrarian movements have proven to be influential networks for advocating politics, effectively mobilizing support, influencing policy, and drawing global attention to the issues peasants and small farmers face. However, more research is needed regarding everyday peasant politics. This gap involves examining the connections between land struggles and peasant politics, especially the agency and resistance strategies peasants use to advocate for their rights and challenge oppressive land tenure systems. Understanding the various tactics peasants employ to resist land dispossession and assert their demands for agrarian reform is crucial for informing empowerment initiatives and social justice advocacy (Borras and Franco 2023). Research indicates that regional factors play a key role in shaping the outcomes of agricultural reform cases and influencing the dynamics between landowners and other stakeholders involved in land disputes (Karaan 2021). However, research in the Philippines has focused on issues in Central Luzon (Isaac et al. 2017; Kerkvliet 2002) and Visayas (Bejeno 2021; Côté 2010), while research in Mindanao (Borras and Franco 2005; Milan 2006) needs to be updated. Therefore, there is a need to update the research by exploring the daily, localized actions and negotiations of peasants, which can provide deeper insights into the mechanisms of farmer empowerment and the practical ways in which they navigate and transform their socio-economic landscapes.

This chapter examines the impact of socio-political differences on the strategies of a peasant movement for farmer empowerment in the Philippines and its effect on small-scale agrarian communities. Section 2 provides an overview of peasant mobilization and agrarian movements in the Philippines, focusing on three organizations connected to La Via Campesina (LVC): the Kilusang Magbubukid ng Pilipinas (KMP), the Democratic Kilusang Magbubukid ng Pilipinas (DKMP), and the National Coordination of Autonomous Rural People's Organizations (UNORKA), shedding light on their internal dynamics and strategies. Section 3 traces UNORKA's evolution into a development partner, illustrating how peasant movements adapt their strategies in response to internal and external challenges. Section 4 analyzes the outcomes of UNORKA's campaign for redistributive agrarian reform at the grassroots level, highlighting the impact of socio-political dynamics on farmers' ability to claim their land rights and using Côté's (2010) method to measure contentiousness. The results reveal that socio-political dynamics substantially impact farmers' ability to claim their land rights, with local power structures influencing decision-making processes related to agrarian reform. Additionally, it highlights the role of UNORKA in facilitating successful land transfers. The chapter concludes by discussing the need for informed policy interventions to better address the needs of small-scale agrarian communities.

## 2 Peasant Mobilization and Agrarian Movements: LVC and the Philippines

Although the terms "peasants" and "small farmers" both refer to people involved in agricultural activities, it is important to understand their specific socio-political implications. Peasants typically engage in subsistence farming, producing food primarily for household consumption. While they may grow cash crops, their focus is on survival rather than commercial gain. They often have insecure land rights and may work as tenants, sharecroppers, or laborers on larger estates, which indicates a complex relationship with the land and external power structures. Their livelihoods are closely tied to traditional rural communities, where cultural practices and social structures play a significant role, leading to challenges related to land access, economic security, and social marginalization.

In contrast, small farmers are more commercially oriented, focusing on generating income from their agricultural activities. They may own or lease the land they cultivate, granting them greater autonomy and control over farming decisions. Small farmers are often more integrated into formal markets and prioritize market-oriented production to sell their produce for profit. While they may also be part of rural communities, their livelihood strategies typically revolve around individual or family-based agricultural activities, with a stronger emphasis on farm production and market engagement (Edelman 2013).

Land-related issues have been a central issue to peasant movements throughout history. Struggles over land rights, access to resources, and agrarian reform shape the mobilization efforts of rural communities seeking social justice (Huizer 2001). Through this principle, LVC challenged the dominant narrative that prioritized market mechanisms over social welfare and community well-being and established alliances with several peasant movements engaged in land reforms, including Brazil's Landless Workers' Movement (MST), Mexico's Zapatista movement and the National Coordination of Autonomous Local Rural People's Organizations (UNORKA) in the Philippines et al. (Borras and Franco 2009; Rosset et al. 2006).

The discourse on land issues traditionally focused on market-oriented policies and economic efficiency, often neglecting social justice, human rights, and environmental sustainability. LVC reframed this discourse by introducing the concept of "rights-based land reform", which emphasizes agrarian reform as a human rights obligation and opposes market-driven land policies in favor of genuine government-led reforms. LVC and other civil society organizations reinforce this concept in a statement reflected in the Final Declaration of the World Forum on Food Sovereignty, held in Havana, Cuba, on September 7, 2001.

> Agrarian reform, above all, should be recognized as an obligation of national governments … within the framework of human rights and as an efficient public policy to combat poverty. These agrarian reform processes must be controlled by peasant organizations … guarantee both individual and collective rights of producers over shared lands, and be articulated within coherent agricultural and trade policies. We oppose the policies and programs for the commercialization of land promoted by the World Bank instead of true agrarian reforms by governments. (World Forum on Food Sovereignty 2001)

LVC has positioned their land rights campaign within the universal human rights framework, embedding it in their global campaign for a "Peasants' Charter." It also actively engaged with intergovernmental institutions such as the UN Commission on Human Rights and other (inter)governmental institutions to promote its agenda for land redistribution and social justice (Borras and Franco 2009). This highlights LVC's significant role in influencing global conversations about land redistribution. However, it is important to acknowledge that land issues and priorities for agrarian reform vary greatly across different regions and countries. LVC members have diverse experiences with land grabbing, land rights, and agricultural policies influenced by their specific regional contexts.

Like any large coalition or movement, LVC may encounter internal dynamics, power struggles, and disagreements among member organizations. Variations in organizational structures, leadership styles, and decision-making processes can affect the movement's ability to present a united front on land reform issues (Borras and Franco 2009; Franco 2008). Some organizations within LVC, like the KMP, reject the "rights of peasants" frame because they consider it a Western and neoliberal concept that does not adequately address the root causes of poverty and inequality in the Philippines (Daphi et al. 2022).

## 2.1 Peasant Movements in the Philippines and Their Relationship with LVC

Historically, research on peasant politics has focused on the struggles of rural communities against oppressive landowners, colonial powers, and exploitative economic systems. Peasants are typically portrayed as subaltern groups resisting subjugation and fighting for land rights, fair wages, and social justice. Historical studies highlighted peasants' uprisings, agrarian movements, and their role in shaping political landscapes through collective action and resistance. In contemporary research, peasant politics are studied in the context of rural development, transnational agrarian movements, and the empowerment of small-scale farmers. Peasant politics involve the collective actions, advocacy efforts, and resistance strategies of peasants against challenges such as land grabbing, market monopolies, environmental degradation, and social injustices. Researchers analyze how they mobilize, organize, and negotiate with various stakeholders to assert their rights, improve their livelihoods, and shape agricultural policies and practices (Van Der Ploeg 2010). A notable example of how transnational agrarian movements engage in peasant politics to facilitate empowerment is when LVC, a peasants' organization with more than 200 million members, successfully influenced the United Nations General Assembly to adopt the Declaration on the Rights of Peasants and Other People Working in Rural Areas (UNDROP) in 2018.

The adoption of UNDROP marks a significant milestone for farmer empowerment and clearly demonstrates peasants' capacity to influence decision-making processes and power to shape their livelihood trajectories (Hoddy 2021). UNDROP established a global legal framework to safeguard the rights of peasants and rural workers. It acknowledges peasants' rights to resources like land, seeds, biodiversity, and a safe working environment, thereby legitimizing their claims and providing recourse to legal mechanisms to defend their rights and hold governments and other actors accountable for violations. Through political activism, peasants amplify their voices, advocate for policy reforms, and promote sustainable agricultural practices, prioritizing community well-being and environmental stewardship (Gradoni and Pasquet 2022). As such, TAMs are an expression of peasant politics that provide peasant movements with advocacy platforms, resources, and solidarity needed to amplify their demands and influence policy. On the other hand, peasant movements contribute grassroots insights and energy that drive the global advocacy efforts of TAMs. This synergistic relationship enhances peasants' ability to secure their rights, promotes sustainable and equitable rural development, and underscores their important role in shaping the discourse surrounding rural development and farmer empowerment.

## 2.2   Differing Views of Land Reform: "Rights-Based" Versus "Genuine" Land Reform

Peasant movements in the Philippines have a long history from the Spanish colonial era. These movements are characterized by cycles of collective actions for various types of agrarian reforms, reflecting farmers' persistent struggle for their rights and national land distribution (Borras et al. 2013; Feranil 2005; Lahiff et al. 2013; Riedinger 2018; Sturtevant 1976). Peasant movements have played a vital role in empowering farmers to resist land grabbing and confront oppressive systems of land ownership and control. Within these movements, various forms of empowerment have emerged, fostering collective agency, political consciousness, and peasant mobilization for social justice (Borras and Franco 2023; Putzel 1992; Milan 2006).

Historically, LVC was associated with three agrarian movements in the Philippines: KMP, DKMP, and UNORKA (Borras 2007). Only KMP and Unity Toward Effective Agrarian Reform and Rural Development (PARAGOS) remain official members (Via Campesina 2018) While both LVC and KMP aim to address land issues and promote social justice in agriculture, they differ in their strategic approaches. LVC emphasizes a rights-based framework, focusing on legal recognition and protection of land rights. It considers secure and equitable access to land a fundamental human right. Meanwhile, KMP prioritizes concrete redistributive measures as part of what they term "genuine" land reform.

KMP's approach emphasizes tangible changes in land ownership and distribution, which focuses on legal frameworks and rights recognition under The Comprehensive Agrarian Reform Program (CARP). The "genuine" land reform advocated by KMP aims to address historical injustices and inequities by redistributing land more directly to landless farmers and agricultural workers. KMP emphasizes the redistribution of large landholdings to small-scale farmers to achieve social justice and economic empowerment for marginalized rural communities.

In the context of CARP, social justice emphasizes the need for the fair and just distribution of land, equal opportunities for landless farmers, and protection of the rights of agricultural workers (Monsod 2014). The Department of Agrarian Reform (DAR) implements CARP, which aims to promote social justice by redistributing agricultural lands to landless farmers and farmworkers. DAR's role includes identifying, acquiring, and distributing agricultural lands to agrarian reform beneficiaries (ARBs), providing support services and infrastructure development to ARBs, and addressing land ownership and tenurial rights conflicts. DAR also collaborates with other government agencies to provide livelihood opportunities and promote rural development (Magno-Ballesteros et al. 2017). Over the past three decades, CARP has distributed 4.8 million hectares to almost 3 million beneficiaries (World Bank 2020). However, when it was first enacted in June 1988, CARP was rejected by nearly all peasant organizations. This rejection was based on the key issues of reform coverage and landlord compensation, which the peasant groups considered too far from "genuine" land reform. The KMP rejected CARP, criticizing the policy as "pro-landlord" and "anti-peasant" (Borras 2007, p. 220).

Although KMP is formally recognized, it has been politically marginalized within LVC and has used its continuous membership to hinder UNORKA's inclusion in the network. UNORKA was eventually admitted to LVC as a "candidate member" in 2008 after participating in important global gatherings as an observer (Borras 2010).

Daphi et al. (2022) attributed the declining political influence of KMP to ideological differences. LVC has formed critical yet collaborative partnerships with select factions within the FAO. However, KMP is skeptical about engaging with international organizations, especially those promoting neoliberal interests. In contrast, KMP has historical ties to the radical leftist activism of the Communist Party of the Philippines (CPP)—New People's Army. Its focus on class-based struggle may lead to skepticism regarding the peasants' rights framework, which they perceive as potentially depoliticizing or co-opting their cause. Moreover, KMP's emphasis on "genuine" agrarian reform may lead them to prioritize national issues over international cooperation, especially if they perceive such collaboration as potentially jeopardizing their fundamental goals. This ideological stance and historical context may contribute to its reluctance to engage with international organizations.

Furthermore, KMP's opposition to UNORKA's participation in LVC is also rooted in socio-political disparities (Borras 2010). UNORKA operates with a decentralized and polycentric structure, where local member groups act as centers of power. This setup allows for a wide range of local initiatives and activities. UNORKA uses "small but dramatic" peasant actions to attract media attention, such as occupying government offices and carrying out symbolic demonstrations to highlight the slow progress of land reform cases (Borras 2007 p. 256). The organization actively participates in the struggle for land redistribution on a significant scale, involving about 200,000 hectares and 500 legal disputes affecting at least 90,000 landless rural poor households. This extensive involvement shows the organization's deep commitment to addressing land reform issues at a grassroots level (Borras 2007).

On the other hand, KMP has a more centralized leadership and coordination. It employs media-centric political agitation campaigns and large-scale peasant mobilizations to challenge the existing state of affairs and advocate for systemic change. The organization has been involved in forming coalitions with other peasant organizations and progressive groups to amplify the peasant's voice and advocate for more progressive land reform policies. However, KMP's involvement in peasant mobilization is less extensive than UNORKA's (Borras 2007, p. 46). KMP is deliberately working to impede the growth and influence of LVC in the Philippines due to differences in ideology. This division can potentially weaken the peasant movement's overall effectiveness by diluting efforts and creating rivalries.

## 2.3  Difference in Advocacy Politics Among Peasant Groups in the Philippines

Borras (2007, 2010) chronicled the emergence of early peasant movements in the Philippines in the 1970s and the subsequent organizational split in the early 1990s of the KMP and DKMP. DKMP was formed by key national leaders who chose to dissociate themselves from KMP due to ideological, political, and organizational differences. DKMP aimed to retain the militant tradition of the KMP while taking a more populist position in terms of land reform by advocating for small family farms. However, it ultimately failed to rally and consolidate its forces, leading to a significant reduction in its membership base (Borras 2010, p. 784).

By the second half of the 1990s, most DKMP member organizations had slowly drifted away from their national federation because they became deeply interested in a non-party social movement framework, incorporating key concepts such as autonomy, internal democracy, leadership-membership accountability, and polycentric leadership. This loose political community later led to the official formation of

UNORKA in 2000. By 2003, UNORKA expanded its political influence on regions where major land reform "battles" were being fought, including Central Luzon, the Bicol region, Negros Island, and the Davao region (Borras 2007 p. 233).

In summary, the distinctive organizational and political characteristics of KMP and UNORKA reflect their strategies and priorities in pursuing genuine land reform in the Philippines, influencing the relationships and interactions within the LVC. However, these organizations are vulnerable to internal conflicts, which often lead to organizational splitting and weaken the political influence of the social movement. How social movements restructure their campaigns after splitting and their impact on small-scale agrarian communities remain relatively unexplored.

## 3 UNORKA's Campaign for Redistributive Land Reform

This section outlines UNORKA's campaign for social justice from the perspective of small-scale banana farmers in Mindanao. The initial phase spans from 2000 to 2010, during which they played a prominent role as leading national reform campaigners using the bibingka strategy. From 2010 to 2014, the second phase marked a shift as external funding ceased, prompting peasant leaders and small farmer groups in Mindanao to focus on rebuilding trust within the community and realigning their grassroots strategies. Finally, from 2015 to the present, the third phase is the revival of UNORKA-Mindanao, which is witnessing a redirection of initiatives toward capacity-building and rural development projects in Mindanao. These three phases in UNORKA's trajectory indicate how internal conflict reshapes strategies for farmer empowerment.

### 3.1 More Than just Advocacy: UNORKA's Bibingka Strategy

The "bibingka" is a ricecake cooked in an oven with heat coming from both the top and the bottom, which metaphorically represents a strategy emphasizing a symbiotic interaction between state reformists "from above" and autonomous rural social movements "from below" for successful land redistribution (Borras 1998). Strategies from below, including collective action, legal action, direct land occupation, and networking, aim to assert rural social movements' claims to land, pressure the government for redistributive land reform, and attract media attention to raise awareness and public support. The bibingka strategy integrates direct action, community organizing, and protests with formal negotiations, lobbying, and policy advocacy. It applies pressure from both grassroots and formal political levels to achieve comprehensive and sustainable social change. This strategy engages and empowers local communities to participate in mass protests, direct actions, and regional governance while maintaining continuous pressure on political actors at all levels. This holistic approach addresses systemic issues like inequitable land distribution and social injustices, particularly prevalent in the Philippines due to unresolved land tenure conflicts (Isaac et al. 2017).

UNORKA employed the bibingka strategy in their national campaign from the early 2000s and has advanced massive landholdings under CARP in 30 provinces, notably in the Bondoc Peninsula (Isaac et al. 2017), Masbate, Iloilo (Bejeno 2021), Cebu (Côté 2010), and Davao del Norte (Borras and Franco 2005). This strategy was very effective during Ramos's presidency (1992–1998) under the progressive leadership of DAR secretary Garilao, but since Arroyo's term from 2001 to 2010, the political opportunity for land reform has significantly decreased (Côté 2010).

Franco and Abinales (2007) investigated the challenging conditions during Gloria Macapagal-Arroyo's presidency, which greatly affected UNORKA's campaign due to political persecution and the government's "all-out war" against communist insurgents. This period witnessed the deaths of numerous activists and the criminalization of agrarian reform cases. Judicial courts and their agents were discouraged from handling agrarian reform-related cases. This was part of a larger assault against leftist political activists, creating an atmosphere of fear and insecurity for peasants and activists. The militarization and political persecution created an environment of intimidation and threats for UNORKA members and leaders. UNORKA members and activists from allied organizations became targets of violence and extrajudicial killings during this period. The report lists 11 peasant leaders killed from 1998 to 2006, including the assassination of Enrico "Ka Eric" Cabanit, UNORKA Secretary General, who was murdered in Davao City in 2006. Ka Eric's murder could be linked to the resistance faced by peasant leaders from powerful landlords, their employees, and allies within the state. The push for agrarian reform and land redistribution posed a threat to vested interests, leading to violent reprisals against leaders like Ka Eric, highlighting the dangers faced by those advocating for peasant rights. Karaan's (2021) analysis of CARP cases from 2010 to 2018 proves that the period marked by policy shifts and changes in administration has influenced the significance of Davao in the context of agrarian reform cases.

## 3.2   UNORKA's Campaign for Land Rights in Davao del Norte

UNORKA-Mindanao is a federation of 153 independent ARB organizations operating in all regions of Mindanao. It has been organizing farmer groups in banana-growing areas since 1999. This section delineates the three distinct phases of UNORKA's campaign for redistributive land reform. The initial phase spans from 2000 to 2010, when they held a prominent role as leading national reform campaigners. The second phase, from 2011 to 2014, marked a shift as external funding ceased, prompting a strategic focus on rebuilding trust within the grassroots movement. Finally, the third phase, from 2015 to the present, witnesses a redirection of initiatives toward capacity-building and rural development-oriented projects in Mindanao. These three phases in UNORKA's trajectory serve as evidence of the diversity within Transnational Agrarian Movements (TAMs).

In the first phase of its activities from 2000 to 2010, UNORKA-Philippines was initially registered as the National Coordination of Autonomous Local Rural People's Organizations. Serving as an umbrella organization for a significant portion of the Philippines' peasant movement, UNORKA's primary support base consisted mainly of landless peasants and rural laborers. UNORKA strategically employed state land reform law both as the institutional framework for its activities and as the focal point of its campaign (Franco 2008). Teaming up with the PEACE Foundation network, UNORKA bolstered the national land reform campaign with funding support from the Dutch NGO ICCO. By 2001, UNORKA was directly engaged in the struggle for land redistribution of approximately 200,000 ha and involved in 500 legal disputes affecting at least 90,000 landless rural poor households. By 2003, the UNORKA-PEACE network had extended its organizational and political influence across virtually all of Luzon, the central Philippines, and the plantation belt of Davao-Cotabato (Borras 2007).

Around 2006, concerns emerged among rural activists as NGOs shifted their support from redistributive land reform campaigns to development-oriented projects. Despite these apprehensions, Borras (2007) remained confident that UNORKA's vertical and horizontal connections with other independent peasant movements and their allies would endure, even without external funding. Moreover, internal conflict in UNORKA arose when its leaders adopted different positions on how to deal with DAR during the Arroyo administration. This ultimately led to a UNORKA split in 2009.

In many dialogues and actions, the groups intended to resolve the land cases and advance the recognition of their rights. However, both parties ended up disagreeing and becoming emotional, which "soured" the relationships between the peasant organizations, NGOs, and DAR and between groups and individuals (Bejeno 2021, p. 79). As Bonita, a UNORKA coordinator, shared: "I think the struggles of UNORKA-Philippines lasted only until 2010. We were neglected after the funding support from NGOs stopped. So, we wanted a clean slate when we organized the Federation and started conceptualizing the paradigm shift in 2014."

In the second phase, from 2011 to 2014, local allies operating in the banana plantations in Davao aim to revitalize the social movement by focusing on the specific regional issues they face in Mindanao. Despite their efforts, many small farmer groups remain suspicious of the leaders, with grievances regarding representation emerging. Some farmers felt betrayed by their national counterparts and NGO partners, who they believed were colluding with the banana companies. It took a year and countless discussions and negotiations for UNORKA-Mindanao to move toward a new direction. Ronita talked about this episode as follows:

> We could not sustain the campaign. Those NGOs who helped us are two-faced. They are in cahoots with the banana companies. For UNORKA, we believe in the principle of Social

> Justice of CARP, which affirms our control of CLOA-awarded lands, including the decision-making. There are people from the NGOs who claim to represent ARBs and yet, directly transact with the companies and all expenses care of the cooperatives. They also claim professional fees from the companies/buyers. ARB's members of coops discovered it.

Finally, in 2015, the leaders of the Mindanao organizations officially registered UNORKA as "Kahugpungan sa Gawasnong mga Organisasyon sa Kabanikanhang Mindanao," marking the third phase. Retaining its Visayan name symbolizes the organization's commitment to addressing the needs of small-scale farmers in Mindanao. However, UNORKA members now face new challenges, particularly in legitimizing their claims to land.

The transfer of land titles has been delayed due to legal disputes, making it difficult for farmers to access and cultivate the land awarded to them. Landowners, who have greater financial resources and legal leverage, have shifted the balance of power in their favor, challenging ARB's ability to assert the rights of the farmers. High litigation costs deter farmers from pursuing legal actions, creating barriers to justice and prolonging the resolution of land disputes.

Furthermore, flaws in the CARP implementation, bureaucratic inefficiencies, and lack of enforcement mechanisms allow vested interests to manipulate the system, undermining the intended benefits for farmers. The shift to online hearings during the COVID-19 pandemic has exacerbated delays in resolving agrarian reform cases, with decisions being postponed under the guise of pandemic-related issues. These delays consume time and resources, frustrating peasants. UNORKA currently faces legal, structural, economic, and political challenges that undermine the program's objectives and perpetuate inequalities in land ownership and access to resources. Addressing these systemic issues is crucial for advancing genuine agrarian reform and empowering farmers to secure their rights to land and livelihood.

### 3.3   UNORKA's Evolution as Partners for Development

UNORKA has taken significant steps to address these issues by strengthening its relationships with local allies through several key strategies. First, they cultivate a spirit of volunteerism among their members and peasant leaders, fostering active participation and community engagement. Second, they have secured legal partnerships with pro-reform lawyers like Jamie Lopoz, who assist in navigating legal battles to protect their rights. Third, UNORKA has sought out NGO-supported livelihood programs to enhance the capabilities of farmer organization leaders and support their pursuit of self-determination. The Mindanao Development Authority (Minda) recognized UNORKA's two decades of campaign experience, acknowledging its long-standing commitment. This recognition led to a fruitful partnership with the Center for Asian Mission for the Poor (CAMP), which secured support from the Government of Korea. As a result, livelihood projects were successfully launched on March 13, 2023, marking a significant milestone in UNORKA's journey. Additionally, they have redirected their advocacy efforts by scheduling dialogues with the Department of Agrarian Reform (DAR) to voice the concerns of ARBs. Finally, they leverage their position as the sectoral representative for Farmers and Landless Rural Workers in the National Anti-Poverty Commission (NAPC) to advocate for peasants' rights and livelihoods in official political arenas.

According to Daphi et al. (2022), social movements initially form alliances around shared goals, interests, and resources. Over time, internal conflicts, negotiations, and external challenges arise, shaping the organization's structure and culture. In response, coalitions may consolidate their efforts, expand their reach, or deepen their impact. If these challenges become overwhelming, dissolution might be considered, allowing members to pursue other partnerships. The resurgence of UNORKA-Mindanao in 2015 is a testament to the resilience and determination of our peasant leaders and farmer groups. After disbanding, they chose to renew their alliance and localize activities. This strategic shift underscores their continued dedication to fortifying local alliances and effectively addressing community-specific issues.

In conclusion, addressing the systemic issues that undermine agrarian reform and perpetuate inequalities in land ownership is crucial. UNORKA-Mindanao's efforts to build strong local alliances, secure legal support, and enhance the capabilities of farmer leaders are vital steps toward advancing genuine agrarian reform and empowering farmers to secure their rights to land and livelihood. The next section will discuss the extent to which these strategies have impacted farmers.

# 4  Farmer Empowerment Strategies in Davao del Norte

This section explores how land ownership empowers farmers and influences their politics in securing their rights to land and resources. Côté (2010) highlights the significance of evaluating landowner resistance for successful land redistribution. By analyzing the outcomes of UNORKA's 20-year efforts to help peasants claim land rights in Mindanao, we uncover the substantial impact of local power structures and the resistance from banana elites on agrarian reform outcomes.

## 4.1  Analytical Framework for Contentious Land Transfers

Côté's (2010) research noted that structural factors such as crop type and land tenure status can impact the power dynamics between peasants and landowners in rural areas. These factors play a crucial role in determining the level of resistance from landowners towards the processes involved in the Comprehensive Agrarian Reform Program (CARP). The key processes affected by these structural features include land coverage, which determines which lands are eligible for reform; the identification of Agrarian Reform Beneficiaries (ARBs), which involves selecting the beneficiaries entitled to land; land compensation, which establishes the compensation for landowners; and the installation of ARBs, which facilitates the transfer of land to the beneficiaries.

Côté categorizes the level of contentiousness based on the resistance shown by landowners towards the CARP processes. Highly contentious settings, where there is strong resistance from landowners to land reform processes, lead to challenges for farmers in securing their land rights. This resistance can result in delays in the implementation of land reform, causing farmers to receive less land than they are entitled to and experiencing long delays before being installed as ARBs. In moderately contentious settings, where there is resistance to only two of the four processes, farmers may have a better chance of claiming their land rights compared to highly contentious settings. Farmers have a significantly higher likelihood of successfully claiming their land rights in somewhat contentious settings, where there is resistance to only one or none of the processes. Therefore, the level of contentiousness in a structural setting significantly influences the outcome of land reform implementation and farmers' ability to assert their land rights.

## 4.2 Landowner Resistance in Davao del Norte

Interactions and conflicts between landowners and beneficiaries are crucial in shaping the outcomes of legal disputes related to agrarian reform and land distribution, especially in Davao (Karaan 2021). In Davao del Norte, banana production is predominantly controlled by four influential families and three companies: Soriano (22.49%), Floriendo (12.50%), Lorenzo (15.81%), Dizon (2.12%), Dole (32.58%), Tristar Group (4.15%), and Marsman-Drysdale Group (4.6%). These groups have established dominance through historical relationships and political structures (De los Reyes and Pelupessy 2009). The cases of farmer conflict and landowner resistance are summarized in Table 1.

The Lorenzos achieved a notable accomplishment by successfully excluding all their banana plantations from land distribution, covering approximately 7000 ha and affecting around 7000 worker households across the Davao peninsula (Borras and Franco 2005). Among the farmer groups significantly impacted by this exclusion was the Gawasnong Maguuma Agrarian Reform Beneficiaries Association Inc. (GMARBAI), originally part of the Hijo Agrarian Reform Beneficiaries Cooperative (HARBCO). HARBCO, which received a collective Certificate of Land Ownership Award (CLOA) for 579 ha of land, entered into a 10-year Agri-Ventures Agreement (AVA) with HPI in December 1998, which later assigned its rights to Lapanday Foods Corporation (LFC), owned and operated by the Lorenzo family.

However, the AVA contained numerous provisions heavily favoring LFC at the expense of the cooperatives' autonomy and fair treatment. The AVA stipulates, among others, that the LFC set the buying price for bananas produced for the Japanese market at $2.10 per 13 kg (kg) net box; the LFC shall have the right to handle the operation of the cooperatives' farms if in LFC's "opinion" that the "success of the crop is endangered" due to the cooperatives' failure to follow LFC "prescribed cultural

**Table 1** Landowner resistance and degree of contentiousness

| Farmer organization | Company | Landowner resistance to the process of | | | | Structural setting |
| --- | --- | --- | --- | --- | --- | --- |
| | | Land coverage | Land compensation | Identification of ARBs | Installation of ARBs | |
| GMARBAI | Lapanday Foods Corporation | Strong | Strong | Moderate | Strong | Highly contentious |
| SIFARBCO | Marshman-Drysdale Group | Strong | Strong | Weak | Strong | Highly contentious |
| PARBEMCO | JK Mercado & Sons | Weak | Weak | Weak | Strong | Somewhat contentious |

practices." GMARBAI challenged the AVA due to its disadvantageous provisions, including LFC's authority to set banana prices and assume farm operations based on subjective criteria. Despite GMARBAI members' refusal to sign the agreement, LFC charged the cooperative's leaders and members with contract violation, citing the collective CLOA requirement for agreement.

Conflict arose when some HARBCO officials and LFC conspired to take full control of banana operations legally, leading GMARBAI to feel forcibly removed from their farm. LFC hired a private army and posted threatening signs along the fence warning, "Intruders will be shot, survivors will be shot again," and refused to recognize GMARBAI's legal status. In response, UNORKA-Mindanao supported GMARBAI in land occupation in 2009 and 2016; they successfully secured 23 ha out of the 85.6 ha they are entitled to. However, the violent encounter resulted in seven wounded and the unfortunate deaths of two of their comrades. Despite legal efforts, a Compromise Agreement affirmed LFC's rights over the disputed land, with the Regional Trial Court of Davao City reiterating the finality and validity of the Compromise Agreement in April 2017. Despite lacking supporting documents, LFC alleged that GMARBAI owed them 14 million pesos for unpaid benefits and dividends. Furthermore, LFC declared they would only recognize GMARBAI's legal status and install them once this purported legal obligation was settled. Thus, GMARBAI once again filed a motion for reconsideration in 2021 to dispute the 14 million charge and urge the court to legally turn over the rights to the portion of land allocated for GMARBAI.

In September 2000, Marsman Estate Plantation, Inc. (MEPI) generously donated 800 hectares of CARP-covered land to 762 eligible ARBs in Sto. Tomas, Davao del Norte, relieving the government of land acquisition costs and the ARBs of repayment obligations. Under the agreement, MEPI would lease the land back from the ARBs for 30 years, paying them rent and providing generous benefits, including high wages, health, education, and financial assistance. However, MEPI cunningly structured the compensation as a cash advance, deducting payment packages from farm profits. SIFARBCO signed the lease contract with MEPI, but in 2002, they discovered their CLOAs were collectively owned due to MEPI's refusal to follow individual partition guidelines, sparking suspicions of collusion between MEPI and the Provincial Agrarian Reform Coordinating Committee (PARCCOM) and other local DAR employees. MEPI's use of the land as collateral for a 1.2 billion loan from the International Finance Corporation (IFC) without farmer consultation further exacerbated tensions. In 2007, although individual CLOAs were signed, MEPI resisted awarding them, opposing the growership model preferred by farmers for land autonomy. Deadlocked negotiations ensued as MEPI refused to release farmers from the leaseback agreement, prompting SIFARBCO and UNORKA-Mindanao to initiate a land occupation to assert their rights. MEPI responded with violent dispersal tactics, injuring one member with a gunshot wound and jailing 13 ARBs for 40 days.

In 2011, SIFARBCO elevated its concerns to the Presidential Agrarian Reform Council (PARC); however, it still needs to address its concerns. MEPI instead initiated an increase in payment for land rentals to appease the farmers. Nevertheless, SIFARBCO wants to "till their land in accordance with the spirit of land reform."

Finally, in 2016, PARC convened and unanimously agreed on the annulment of the lease contract agreement with MEPI. However, MEPI filed a motion for reconsideration and, at the same time, initiated a modified contract about land rentals and other benefits to lure farmers to sign a new contract. Some farmers were convinced, but SIFARBCO remained steadfast and waited for the final decision on the annulment. Finally, after a series of follow-ups, the PARC convened in 2021; however, it was presided over by PARC vice chairman John Castriciones, who was DAR secretary at that time. In a most unfortunate turn of events, PARC released its decision unanimously, granting MEPI the motion for reconsideration and diverting the annulment of the old contract in favor of the new contract issued by MEPI.

PARBEMCO's experience differed from that of GMARBAI and SIFARBCO. Their landowner, JK Mercado & Sons, was supportive of agrarian reform and believed that farm workers should have the right to own their land. The landowner's son even became an ally to the farmers. Despite opposition from some family members, he encouraged ARBs to assert their legal rights to the land. The ARBs were able to claim their land smoothly, with the only opposition coming from other children of the landowner who were fighting over inheritance. UNORKA-Mindanao successfully assisted PARBEMCO in defending their rights against inheritance claims.

## 4.3  Outcomes of Peasant Mobilization and Peasant Politics

The experiences of peasant farmers in Davao del Norte illustrate the intricate interplay between local socio-political dynamics and farmer empowerment. As highlighted by Carranza (2015), conflicts between pro- and anti-agrarian reform factions often lead to violence against land rights claimants, exposing the vulnerability of farmers asserting their rights. Moreover, the government's failure to protect these fundamental human rights exacerbates the challenges farmers face in securing their land and livelihoods.

Table 2 further underscores the disparities in resistance strategies employed by landowners within the land redistribution system despite the shared circumstances of cultivating the same crop and being farmworkers. These insights emphasize the pivotal role of peasant politics in shaping the access and control small farmers have over land resources. Thus, addressing these challenges is imperative for genuine agrarian reform and the empowerment of farmers, as discussed in the conclusion.

Peasant farmers in banana farming communities in Davao del Norte engage in everyday peasant politics through various avenues, reflecting their ongoing efforts to address local issues and assert their rights within their communities: (1) Community Organizing: Peasant farmers mobilize collective action around shared concerns through community organizing activities. They update the status of official cases and collaborate to address common challenges collectively. (2) Legal Advocacy: Peasant farmers undertake legal advocacy to challenge injustices, defend their rights, and seek redress through legal channels. Both GMARBAI and SIFARBCO have filed complaints and lawsuits and sought legal assistance from pro-bono lawyers to

**Table 2** Land occupation outcomes

| Farmer organization | Structural setting | Expected outcomes | Actual outcomes |
|---|---|---|---|
| GMARBAI | Highly contentious | **Weak implementation** | **Unimplemented** |
| | | • Installation delay: more than five years<br>• Land received: less than what peasants were entitled to | • Petition to install farmer beneficiaries rejected because of internal dispute<br>• High risk of foreclosure due to unpaid amortization |
| SIFARBCO | Highly contentious | **Weak implementation** | **Partial implementation** |
| | | • Installation delay: more than five years<br>• Land received: less than what peasants were entitled to | • Individual CLOAs signed but not awarded<br>• DAR delayed CLOA awarding and allowed the company to file a motion for reconsideration<br>• PARC En Banc favored the company and approved the motion |
| PARBEMCO | Somewhat contentious | **Full implementation** | **Full implementation** |
| | | • Installation delay: more than five years<br>• Land received: less than what peasants were entitled to | • The owners were pro-ARB and helped the farmers claim their individual CLOAs |

address issues such as land disputes or human rights violations. (3) Economic Initiatives: Peasant farmers engage in economic initiatives like cooperative farming and agroecology practices to enhance their economic resilience and reduce dependence on external actors. For example, SIFARBCO is transitioning to organic banana cultivation and intercropping with corn to promote sustainable farming. Meanwhile, GMARBAI is gradually shifting from collective to individual farming to grant farmers autonomy over production. Additionally, PARBECO, facing crop losses due to fusarium wilt, is diversifying crops by planting okra and other vegetables to allow the land to rest from pesticides and heal the soil.

These everyday efforts contribute to peasants' economic resilience, reduce dependence on external actors, and strengthen their sense of ownership and responsibility towards the land. By engaging in community organizing, legal advocacy, and economic initiatives, peasant farmers in Davao del Norte assert their rights and work towards building sustainable livelihoods within their communities.

## 5  Conclusion

The strategies peasant farmers use in Mindanao demonstrate their nuanced approach to navigating local power dynamics to secure access to and control over land. Initially, they employ advocacy politics through mass mobilization and militant protests at the national level, which helps them gain initial access to land. However, the sustainability of these political engagements is often undermined by financial constraints, necessitating adaptive measures. In response, peasants strengthen local alliances, revise their action plans, and seek new areas of collaboration to address evolving challenges. While these local alliances successfully facilitate access to land, it is crucial to distinguish between mere access and actual control over land. Access enables livelihoods and resource utilization, but true security and autonomy are achieved through control. Without control, individuals remain vulnerable to displacement or exploitation by more powerful actors. Therefore, further research, particularly longitudinal analysis of land access and control dynamics, is essential to assess the long-term impacts of both advocacy and local alliance strategies on land tenure and security.

Despite two decades of campaigning for land reform, small-scale banana farmers have gained initial access to land but struggle to maintain control. Successfully navigating the contentious land distribution system, they now face new obstacles in legal and market strategies. UNORKA's efforts to forge alliances with international development organizations demonstrate ongoing adaptation to shifting socio-political and economic landscapes. However, these efforts require ongoing revision to remain effective.

Overall, this chapter sheds light on the complex interplay between organizational dynamics, strategic adaptation, local empowerment, and policy implications within peasant politics. These insights are invaluable for policymakers and stakeholders engaged in agrarian reform and rural development, offering a deeper understanding of the challenges and opportunities faced by peasant movements. Informed policy interventions can better address the needs and priorities of small-scale agrarian communities, leading to more effective outcomes in rural development initiatives.

## References

Bejeno C (2021) On the frontlines: peasant women and land reform struggles in the Philippines. Doctoral Dissertation, Erasmus University Rotterdam, the Netherlands. https://repub.eur.nl/pub/136981

Borras SM Jr (1998) Bibingka strategy to land reform and implementation: Autonomous peasant mobilizations and state reformists in the Philippines. ISS working papers—general series 19017. International Institute of Social Studies of Erasmus University Rotterdam (ISS), The Hague. https://ideas.repec.org/p/ems/euriss/19017.html

Borras SM Jr (2007) Pro-poor land reform: a critique. University of Ottawa Press

Borras SM Jr, Carranza D, Franco JC (2013) Anti-poverty or anti-poor? The world bank's market-led agrarian reform experiment in the Philippines. In: Market-led agrarian reform. Routledge, pp 141–160

Borras SM Jr (2008) La Vía Campesina and its global campaign for agrarian reform. J Agrar Change 8(2–3):258–289. https://doi.org/10.1111/j.1471-0366.2008.00170.x
Borras SM Jr (2010) The politics of transnational agrarian movements. Dev Change 41(5):771–803. https://doi.org/10.1111/j.1467-7660.2010.01661.x
Borras SM Jr, Edelman M, Kay C (2008) Transnational agrarian movements: origins and politics, campaigns and impact. J Agrar Change 8(2–3):169–204. https://doi.org/10.1111/j.1471-0366.2008.00167.x
Borras SM Jr, Franco JC (2005) Struggles for land and livelihood: redistributive reform in agribusiness plantations in the Philippines. Crit Asian Stud 37(3):331–361. https://doi.org/10.1080/14672710500200383
Borras SM Jr, Franco JC (2009). Transnational agrarian movements struggling for land and citizenship rights. IDS Work Pap 323:1–44. https://doi.org/10.1111/j.2040-0209.2009.00323_2.x
Borras SM Jr, Franco JC (2023) Scholar-activism and land struggles. Practical Action Publishing
Carranza D (2015) Agrarian reform and the difficult road to peace in the Philippine countryside. Norwegian Peacebuilding Resource Center-NOREF. https://reliefweb.int/report/philippines/agrarian-reform-and-difficult-road-peace-philippine-countryside
Côté DJ (2010) Successful strategies for the implementation of land reform: a peasants' account from the Philippines. Master's Thesis, University of Montreal, Canada. https://papyrus.bib.umontreal.ca/xmlui/handle/1866/3888
Daphi P, Anderl F, Deitelhoff N (2022) Bridges or divides? Conflicts and synergies of coalition building across countries and sectors in the global justice movement. Soc Mov Stud 21(1–2):8–24. https://doi.org/10.1080/14742837.2019.1676223
De los Reyes JH, Pelupessy W (2009) Agrarian reform in the Philippine banana chain. Discussion Paper 2009.3. IOB Antwerpen, Belgium. https://ideas.repec.org/p/iob/dpaper/2009003.html
Douwe Van Der Ploeg J (2010) The peasantries of the twenty-first century: the commoditisation debate revisited. J Peasant Stud 37(1):1–30. https://doi.org/10.1080/03066150903498721
Edelman M (2013) What is a peasant? What are peasantries? A briefing paper on issues of definition. First Session of the Intergovernmental Working Group on a United Nations Declaration on the Rights of Peasants and Other People Working in Rural Areas, Geneva, 15–19 July 2013. https://www.ohchr.org/sites/default/files/Documents/HRBodies/HRCouncil/WGPleasants/MarcEdelman.pdf
Edelman M (2024) Peasant politics of the twenty-first century: transnational social movements and agrarian change. Cornell University Press
Feranil SH (2005) Stretching the limits of redistributive reform: lessons and evidence from the Philippines under neoliberalism. In: Moyo S, Yeros P (eds) Reclaiming the land: the resurgence of rural movements in Africa, Asia, and Latin America. Zed Books, London, New York, pp 257–284
Franco JC (2008) Making land rights accessible: social movements and political-legal innovation in the rural Philippines. J Dev Stud 44(7):991–1022. https://doi.org/10.1080/00220380802150763
Franco JC, Abinales PN (2007) Again, they're killing peasants in the Philippines: lawlessness, murder, and impunity. Crit Asian Stud 39(2):315–328. https://doi.org/10.1080/14672710701339501
Gradoni L, Pasquet L (2022) Voice under domination: notes on the making and significance of the United Nations declaration on the rights of peasants. Eur J Int Law 33(1):39–64. https://doi.org/10.1093/ejil/chac014
Hoddy ET (2021) Peasants' rights and agrarian violence in transitional settings: from transitional justice to transformative agrarian justice. J Hum Rights 20(1):91–109. https://doi.org/10.1080/14754835.2020.1850242
Huizer G (2001) Peasant mobilization for land reform: historical case studies and theoretical considerations. In: Ghimire KB (ed) Land reform and peasant livelihoods: the social dynamics of rural poverty and agrarian reforms in developing countries. ITDG Publishing, London, pp 155–198

Isaac F, Carranza D, Aceron J (2017) From the ground up: multi-level accountability politics in land reform in the Philippines. Accountability Research Center. Accountability working paper 2. https://accountabilityresearch.org/publication/from-the-ground-up-multi-level-accoun tability-politics-in-land-reform-in-the-philippines/

Karaan MML (2021) An empirical study of comprehensive agrarian reform program (CARP) cases. Phil LJ 94:401

Kerkvliet BJ (2002) Everyday politics in the Philippines: class and status relations in a Central Luzon village. Rowman & Littlefield

Kerkvliet BJT (2013) Everyday politics in peasant societies (and ours). In: Critical perspectives in rural development studies. Routledge, pp 215–231

Lahiff E, Borras SM Jr, Kay C (2013) Market-led agrarian reform: policies, performance and prospects. In: Market-led agrarian reform: critical perspectives on neoliberal land policies and the urban poor. Routledge, pp 1–20

Magno-Ballesteros M, Ancheta J, Ramos T (2017) The comprehensive agrarian reform program after 30 years: accomplishments and forward options. PIDS discussion paper series. https://pid swebs.pids.gov.ph/CDN/PUBLICATIONS/pidsdps1734.pdf

Milan FMP (2006) The politics, landowners' resistance and peasants' struggle for genuine land reform in the Philippines. Master's thesis. University of Hohenheim, Stuttgart, Germany. https://stiftung-fiat-panis.de/images/DF/DF413.pdf

Monsod CS (2014) Social justice. Ateneo LJ 59(3):691–722

Putzel J (1992) A captive land: the politics of agrarian reform in the Philippines. Catholic Institute for International Relations

Riedinger JM (2018) Everyday elite resistance: redistributive agrarian reform in the Philippines. In: Kannabiran K (ed) The violence within: cultural and political opposition to agrarian reforms. Routledge, pp 181–218

Rigg J (2024) Peasants and the peasantry. In: The Oxford handbook of agricultural history. Oxford University Press, p 129

Rosset P, Patel R, Courville M (2006). Promised land: competing visions of agrarian reform. Food First Books

Scott JC (1985) Weapons of the weak: everyday forms of peasant resistance. Yale University Press

Sturtevant DJ (1976) Popular uprisings in the Philippines 1840–1940. Cornell University Press

Tadem EC (2009) The Filipino peasant in the modern world: tradition, change and resilience. Philipp Polit Sci J 30(1):1–34. https://doi.org/10.1163/2165025x-03001001

Via Campesina (2018) La Via Campesina members. https://viacampesina.org/en/wp-content/upl oads/sites/2/2018/03/List-of-members.pdf

World Bank (2020) Philippines: new project to help provide individual land titles to 750,000 agrarian reform beneficiaries. World Bank Group. https://www.worldbank.org/en/news/press-release/2020/06/26/philippines-new-project-to-help-provide-individual-land-titles-to-750000-agrarian-reform-beneficiaries

World Forum on Food Sovereignty (2001) Final declaration. Havana, September 7. https://www.iatp.org/documents/final-declaration-of-the-world-forum-on-food-sovereignty-0

# Reduce Corruption Risks in Defense Personnel: Government Defense Integrity Index Framework

Yao Chun Tsao

**Abstract** In the Government Defense Integrity Index (GDI) 2020 Taiwan national assessment report, Taiwan overall received a good rating of Band "B" (low corruption risk). Among the five major aspects of the assessment, personnel corruption risk received a Band "A" rating. According to (Pyman in Tackling Defense Corruption. Ethical Dilemmas in the Global Defense Industry, 259, 2023), the government can improve its military integrity using GDI evaluation results to compare with better-performing countries and to improve our relatively weak areas accordingly. In this chapter, I provide specific solutions to the potential risks of corruption among the personnel of the Ministry of National Defense, especially the protection of whistle-blowers in the national military, the verification of important and sensitive officials, and the selection and appointment of middle- and high-level military officials and civil servants. The objectivity of promotions is a key point that can be strengthened in the short term. Based on this, not only can Taiwan strive for good GDI evaluation results in 2025, but it can also improve the substantive level of national defense integrity, which is also the crucial contribution of this chapter.

**Keywords** Military personnel · Corruption risks · Government defense integrity index · Transparency international

The research was funded by the National Science and Technology Council of Taiwan (R.O.C.), grant No: NSTC-2637-H-230-001. Although the author is the GDI advisor to the Ministry of National Defense of Taiwan, this chapter only represents his personal research opinions and does not represent any specific Institution's point of view.

Y. C. Tsao (✉)
Cheng Shiu University, Kaohsiung, Taiwan
e-mail: k0486@gcloud.csu.edu.tw

# 1  Introduction

The Government Defense Integrity (GDI) Index is the world's first authoritative index to evaluate the transparency of national defense and military affairs. Each of the five major aspects—politics, finance, military operations, and defense procurement—has its own thematic questions, with each topic divided into several different sub-sections. The score for each sub-item ranges from 0 to 4. The GDI provides a total of about 1000 patterns, and the DSP aggregates the scores of each aspect of the question, placing the rated country in one of the six assessment scales (A–F). The six levels indicate the level of corruption from the least corrupt to the most corrupt: A (very low corruption), B (low corruption), C (moderate corruption), D (high corruption), E (very high corruption), and F (extreme corruption). The final GDI 2020 results were released at the end of 2021 (Transparency International, Defence and Security 2021), with New Zealand ranked No. 1 in the world, achieving the only Band A rating. Taiwan is ranked 6th in the world, along with eight other countries, including the United Kingdom, Norway, Belgium, the Netherlands, Germany, Switzerland, and Latvia, all receiving a B rating (Tsao 2023).

According to the GDI 2020 Taiwan Assessment Report (TI-DS 2021), Taiwan has been investing heavily in enhancing its self-defense capabilities and implementing a new strategic defense concept based on asymmetric warfare. Taiwan's existing defense governance mechanisms are strong, formal, and resource-rich. The management and ethics frameworks for Taiwan's defense, military, and civilian personnel include strong codes of conduct. Additionally, the personnel payroll system and appointment system have been formalized and systematized, substantially reducing the risk of corruption and abuse of power caused by uneven appointments. However, there is still room for increased transparency in the recruitment and promotion process of defense-related personnel. The Legislative Yuan is not currently involved in reviewing the appointment of senior military personnel. Furthermore, efforts to expose and report fraud still lack substantial protection for whistleblowers. Although Taiwan is currently formulating a whistleblower protection law, it has not yet been completed, so it relies heavily on individual prosecutors and lacks significant institutional protection.

Usually, a country will have a state of natural corruption, but it will not make the necessary improvements without a strong shock from external sources and subsequent media exposure for the government to wake up and intervene (Haywood et al. 2019). Corruption is notoriously difficult to observe, relying mainly on country rankings based on the International Corruption Index (Ceschel and Homberg 2022; Tsao and Hsueh 2023). In Taiwan, the Ministry of Justice and the Ministry of National Defense have attached great importance to the evaluation of the GDI, and in recent years, they have invited external integrity experts to participate in intensive discussions of the evaluation project. This chapter is an in-depth post-evaluation review and analysis of the personnel corruption risk report (one of the author's five major evaluation projects) within the five major evaluation projects. It aims to provide specific suggestions for improvement from the objective perspective of international integrity evaluation, for the reference of Taiwan's anti-corruption units and the Ministry of National Defense.

## 2 Literature Discussion

Więcek-Starczyńska et al. (2019) point out that there is often a high correlation between low corruption and a high degree of globalization, a stable economy, and a high level of human resources. Pyakuryal and Cox (2019) found that the pattern of corruption in Southeast Asia is inconsistent, but common challenges across countries include the cyber shields established by bureaucrats that resist anti-corruption efforts in order to maintain the status quo and entrench a culture of corruption. Corruption is widely regarded as a social plague that affects most countries around the world. In recent years, there has been an increasing focus on combating corruption, not only through anti-corruption policies aimed at punishing corruption but also through preventive policies aimed at limiting the occurrence of potential impacts. While corruption is widespread, prevention strategies need to be adapted to local conditions to effectively reduce it. Although research on the causes, preventive measures, and consequences of corruption has attracted academic attention, the relationship between the causes of corruption and related anti-corruption measures is still rarely discussed (Ceschel and Homberg 2022). If the analysis and understanding of the impact of corruption inspire institutional differentiation in initiating preventive policies rather than just comparing them, "preventing" corruption makes it increasingly important to study the nature and causes of corruption in depth (Dimant and Tosato 2018).

Klitgaard et al. (2000) state that individuals have access to the benefits of corruption depending on the degree of monopoly they have over the service or activity, the discretion they have to decide who should receive how much, and the degree of responsibility for their activities. It can be seen that in terms of opportunities, corruption can be effectively prevented by improving the control system of business operators, limiting their monopoly conditions, and improving the transparency of their behavior. Other arguments focus on the study of incentives to engage in corrupt behavior. Since these are inevitably related to the characteristics of the environment and the organizational context, it is necessary to consider how they interact with the characteristics of the organizational actors, which may underlie their corrupt behavior.

Luk (2012) proposes a three-pronged approach that advocates a combination of ethical leadership, ethical training, and ethical legislation to maintain the integrity of the civil service in Hong Kong. Well-planned staff training can also alleviate corruption. Berman (2015) discusses in detail how human resource management can support anti-corruption efforts. Regarding human resource management systems, several empirical studies have investigated the anti-corruption effects of different human resource practices. Human resource management can also instill professional commitment, loyalty, and pride in serving the public. Thus, highly selective recruitment procedures, appropriate induction and ongoing training (Beeri et al. 2013), and socialization efforts may help build a public sector workforce that is less inclined to engage in corrupt practices. In this sense, a strong human resource management system can serve as an antidote to corruption (Berman 2015). On the same subject, specific personnel management measures were also considered potential corruption prevention tools. For example, under the rotation system, public officials are less likely to accept bribes, as this leads to fewer cases in which public officials' decisions are manipulated (Fišar et al. 2021).

A traditional argument in the corruption literature focuses on the remuneration of public officials. Higher remuneration (compared to private sector remuneration) that allows individuals to maintain their living expenses is often seen as a deterrent to corruption. Individuals who work in such conditions are less inclined to accept bribes or other gifts. While part of the study by van Veldhuizen (2013) confirms this thesis, arguing that higher pay makes individuals less susceptible to corruption, a broader review of the literature reveals a more nuanced picture. However, this study does not provide systematic evidence. In contrast, Chen and Liu (2018) studied bribes received by Chinese officials based on court litigation data and found a U-shaped relationship. According to their findings, when the base pay is low, an increase is effective as an anti-corruption measure, but when the pay is already at a high level and bribery can be negotiated, the increase is not effective.

Navot et al. (2016) examined cross-country data from 58 countries, emphasizing that higher pay may exacerbate public corruption and demonstrating that people with a stronger sense of belonging tend to be less influenced by monetary incentives. This finding is consistent with the theory of intrinsic motivation. Campbell (2020) compared more than 100 countries and found that the effect of pay on corruption was negative when patronage was low, but not significant in countries where patronage was present in the hiring process. Mangafić and Veselinović (2020) analyzed the impact of personal determinants on the likelihood of participating in bribery, confirming that specific personal characteristics can predict corrupt behavior. Additionally, dishonesty is strongly negatively correlated with public service motivation (Olsen et al. 2019). The study by Thaler and Helmig (2016) further highlights the impact of the adoption of a code of conduct or ethical leadership on individual behavior, showing that only ethical leadership can have a positive impact on employees' attitudes related to the organization (Janenova and Knox 2020).

Pyman (2023) specifically points out that Taiwan's military may still have the risk of artificial favoritism in personnel promotion and appointment. Personnel promotions in the military should break down possible biases by ensuring that different sources of reference assessments are considered, using a standardized interview process, and employing predefined metrics or criteria for tasks and assessments. Data analysis is also needed to assess trends and validate internal consistency, and the transparent publication of the results of the selection committee (HRC) can provide future organizational accountability (Holt and Davis 2022). The adoption of consistent assessment and transparent reporting procedures for the selection of different official lists is crucial. Without processes to combat unconscious bias, units and leaders tend to favor officials they know, who look like them, or who have similar backgrounds.

Leaders can take practical steps to promote diversity and mitigate unconscious bias. The officer in charge of selection should consider two or more individuals from underrepresented groups when ranking the list of potential candidates. Candidates should be scored transparently so that objective comparisons can be made. Leveraging personal knowledge, skills, and behaviors (KSB) as well as preferences to use a more flexible career model can adapt to changing family structures and demographics. As the identification of KSB matures, cognitive diversity can be assessed by combining KSB, demographics, education, and task history. In contrast to subjective assessments, data-driven personnel selection can identify the top performers. Leaders responsible for selecting officers should be informed about conducting interviews, developing objective evaluation criteria, and evaluating and selecting soldiers to fill the organizational cognitive diversity gap, because unintentional but systematic favoritism of one group over another does exist in values-based organizations (Joy 2019; US Army Talent Management 2020; Grant 2021).

Whistleblowing plays a crucial role in exposing organizational misconduct and systemic corruption in industry and government (Munro and Kenny 2023). However, about one in five people who speak publicly about organizational misconduct are subject to retaliation (Transparency International Ireland 2017). When people speak openly about wrongdoing, they often find themselves in an organization where a culture of silence and complicity prevails. Being excluded from the group can lead to isolation and cause individuals to give up struggling (Devine and Maassarani 2011). When whistleblowers become outcasts, scapegoats, and pariahs in the group to which they once belonged, blind loyalty takes the place of truth (Sylvester 2017), and cultural disharmony seems unsolvable for future cohesion. Thus, whistleblowers are considered honest in terms of interpersonal interactions but are seen as disloyal, resulting in their punishment (Wharton 2019).

In the debate on whistleblowing policy and proposed changes to the law, the dangers of real-name reporting were acknowledged. Therefore, whistleblowing experts believe that it is important to ensure that anonymous whistleblower disclosures are accepted by organizations and that whistleblowers are legally protected (OECD 2019). For example, the European Parliament "considers that the option of anonymous reporting encourages whistleblowers to share information they would not otherwise share" (European Parliament 2017). This particularly affects women and other groups who find themselves in a minority position in certain circumstances, as women are more likely than men to report misconduct if they remain anonymous (IBA 2021).

For military whistleblowers, the process can be more difficult due to less legal protection and the closed trial system. Active-duty military personnel who report any type of waste, fraud, abuse, discrimination, harassment, bullying, or sexual assault are subject to review by the command they report to. Administrative dismissal is difficult to refute because the opinions and needs of the commander take precedence. Command can counter-point servicemen, especially those of lower rank, and claim that the whistleblower is the problem. They can force them to undergo psychiatric examinations to be used against them, divert them to less-than-ideal assignments, provide poor health reports, remove them from duty safety reviews, and eventually retire them. This means that whistleblowers suffer long-term effects of retaliation, depriving them of access to health care, education benefits, pension or severance pay, and the security clearances needed for future employment (Horton 2020). Therefore, it is all the more important for the government to take remedial measures to assist whistleblowers and punish organizations to prevent multiple forms of retaliation.

Recent literature has highlighted the complementary role of whistleblowers in organizations in anti-corruption initiatives, as anti-corruption measures alone may not be sufficient. Chang et al. (2017) focused on the determinants of whistleblowing, arguing that the support of colleagues and the organization are important prerequisites for individuals to be willing to report. To encourage whistleblowing, both organizations and the legal system need to implement safeguards for the individuals involved. Burlando and Motta (2016) developed an anti-corruption enforcement model with a financial subsidy and legalization policy for whistleblowing, which can play a key anti-corruption role. However, they also point to some significant limitations of such policies, particularly the negative impact on people's trust. Capasso et al. (2019) take a different view of law enforcement, assessing the effectiveness of all types of law enforcement, judicial efficiency, and broader related systems. They find that strengthening systems at the national level is the most powerful deterrent to corruption. In other words, building better alignment with the law and enforcing overall improvements in enforcement is more effective in combating corruption than focusing on individual enforcement (Capasso et al. 2019).

The comprehensive literature discussion on several topics with low scores on the theme of defense personnel corruption risk in Taiwan in GDI 2020 can make the strategic solution of the problem analysis more focused, and better understand how Taiwan should respond to GDI before GDI 2025. Graycar and Monaghan (2015) specifically emphasize that corruption differs between democracies and countries in a state of political and economic development or transition. Therefore, anti-corruption measures should vary according to the development situation of a particular country. Taiwan's specific improvement method is to adopt the evidence from evaluation reports of various countries, conduct in-depth content analysis, and finally put forward specific and feasible improvement strategies according to the literature suggestions and Taiwan's national conditions. According to the analysis of Ceschel and Homberg (2022), important research literature on corruption is quantitative (58%) and qualitative (24%). The third is conceptual or critical research, which accounts for 18%. This chapter is the third type, making an important contribution to the theoretical development and explanatory model of corruption research.

# 3 Research Method

This chapter first reviews the anti-corruption literature related to Taiwan's weaknesses in defense personnel as highlighted in the GDI 2020 report, and then conducts an in-depth review of each identified weakness. Pyman (2023) suggests that the evaluation results of the GDI can be used for comparison with other countries, as the GDI employs quantitative evaluation methods, making it feasible to objectively compare the integrity status of different countries and improve accordingly. Self-improvement can be achieved by focusing on questions with low scores (0, 1, or 2) and skipping those with high scores (e.g., 3 or 4). However, given that Taiwan is already ranked as a B-level country in GDI 2020, where most scores are 3 points or above, improving to an A level requires including questions with a score of 3 in advanced study and analysis to enhance the overall level.

To this end, I will first take Band A countries, specifically New Zealand, as the model. If New Zealand's score is not better than Taiwan's in the subject item under review, I will then consider B-level countries with a full score or better than Taiwan's score (Germany, the United Kingdom, Belgium, the Netherlands, Norway, Switzerland, and Latvia as the main reference countries in the GDI 2020 Global Report). If the items to be improved cannot be benchmarked against Band A or Band B countries, countries with higher scores than Taiwan in specific areas (e.g., Japan and Singapore) with similar national conditions will be used as reference samples. If Taiwan does not reach the A level and no country is better than Taiwan, the model answers of GDI 2020 (TI-DS 2021) will serve as the standard for discussion.

In summary, the improvement method involves adopting an evidence-based orientation from the final evaluation reports of various countries, conducting in-depth content analysis, and proposing concrete and feasible strategies based on exemplary answers, literature suggestions, and Taiwan's national conditions (Tsao 2023). Even though GDI 2020 has established model answers for countries to follow, fully understanding the design connotation of the GDI and participating in the GDI evaluation are essential for truly responding to international evaluation requirements. Ordinary scholars may find it challenging to locate suitable answers through literature discussion alone, and thus cannot effectively assist government assessors of the Ministry of Defense in addressing relevant evaluation questions. Therefore, this paper provides direct suggestions and strategies to the Ministry of Defense based on academic theory and the GDI framework.

# 4 Discussions and Strategies

From Table 1, it is clear that the global average score for personnel corruption risk is 53 points, the highest score among the five evaluation categories. Taiwan is ranked in Band A for this item and has the highest score in the five major categories, with a score of 84 points, indicating that this category has contributed the most to Taiwan's overall evaluation score and demonstrates the international evaluation institutions' recognition of the integrity of Taiwan's national defense personnel.

From Table 2, there are a total of 17 questions on the risk level of corruption in defense personnel, and six items have an evaluation grade below A. The protection of whistleblowers in question 36 is rated as Band D, indicating a relatively weak area. The review of the selection process for high-risk positions in question 37 is rated as Band C. Question 38 examines the completeness of the disclosure of the number of military personnel, and this item is rated as Band B. Question 41 measures whether there is a perfect internal and external mechanism for the selection and appointment of middle- and high-ranking personnel in the Taiwan military. Question 42 deals with the fairness and objectivity of military personnel promotions and is rated as Band B. Question 45 determines whether the command system of the Taiwan military is separated from the salary payment system to avoid malpractice for personal gain.

The following sections will focus on these six relatively imperfect questions, addressing their content and key points, and then comparing them with the existing practices of the Ministry of National Defense of Taiwan. Based on the research procedures of this chapter, corresponding strategies will be proposed.

**Table 1** Taiwan's five corruption items in GDI 2020

| Major items | Band of Taiwan | Taiwan scores | Global average |
| --- | --- | --- | --- |
| *Personnel risks* | *A* | *84* | *53* |
| Political risk | B | 79 | 46 |
| Financial risk | B | 81 | 45 |
| Military operational risks | D | 48 | 16 |
| Procurement risk | C | 58 | 37 |
| Total risk/scores | B | 70 | 39 |

*Note* Band A is scored 83–100, indicating a very low risk of corruption. Band B is 67–82 points, indicating a low risk of corruption. Band C is 50–66 points, indicating a moderate risk of corruption. Band D is 33–49 points, indicating a high risk of corruption. Band E is 17–32 points, indicating a very high risk of corruption. Band F scores 0–16, indicating an extremely high risk of corruption
*Source* Transparency International, Defence and Security, 2021. GDI 2020 Global Report
The significance of italics indicates the rating scores under 4 for further study and analysis to enhance the overall GDI level of Taiwan

**Table 2** Risk level of corruption in Taiwan's defense personnel

| Defense personnel corruption risk item: focus | Band | Score |
|---|---|---|
| Q34: Public commitment to integrity | A | 100 |
| Q35: Disciplinary measures for personnel | A | 88 |
| *Q36: Whistleblowing* | *D* | *42* |
| *Q37: High-risk positions* | *C* | *50* |
| *Q38: Numbers of personnel* | *B* | *67* |
| Q39: Pay rates and allowances | A | 100 |
| Q40: Payment system | A | 100 |
| *Q41: Objective appointments* | *D* | *42* |
| *Q42: Objective promotions* | *B* | *69* |
| Q43: Bribery to avoid conscription | A | 100 |
| Q44: Bribery for preferred postings | A | 100 |
| *Q45: Chains of command and payment* | *B* | *75* |
| Q46: Military code of conduct | A | 100 |
| Q47: Civilian code of conduct | A | 100 |
| Q48: Anticorruption training | A | 100 |
| Q49: Corruption prosecutions | A | 100 |
| Q50: Facilitation payments | A | 92 |
| Defense personnel corruption risk rating/score | B | 79 |

*Note* Band A is scored 83–100, indicating a very low risk of corruption. Band B is 67–82 points, indicating a low risk of corruption. Band C is 50–66 points, indicating a moderate risk of corruption. Band D is 33–49 points, indicating a high risk of corruption. Band E is 17–32 points, indicating a very high risk of corruption. Band F scores 0–16, indicating an extremely high risk of corruption
*Source* Transparency International, Defence and Security, 2021. GDI 2020 Global Report
The significance of italics indicates the rating scores under 4 for further study and analysis to enhance the overall GDI level of Taiwan

## 4.1 Q36: Is Whistleblowing Encouraged by the Government, and are Whistle-Blowers in Military and Defense Ministries Afforded Adequate Protection from Reprisal for Reporting Evidence of Corruption, in Both Law and Practice?

### Current Practice

First of all, from the perspective of legal provisions, the "Whistleblower Protection Law" is an important special law for Taiwan to implement its anti-corruption goals, and it is also an important international index to measure the degree of integrity of a country. During the 2022 international review meeting of the United Nations Convention against Corruption, the review committee pointed out in the review report that a law on the protection of whistleblowers should be enacted as soon as possible. At present, Taiwan's Ministry of Defense does have many platforms for officers and men to report corruption and lawlessness, and Taiwan's Ministry of Defense and its subordinate agencies also have their own channels for reporting, which are not shared with other government departments. Taiwan received the lowest score (only 1 point) in this category, which means that the assessor believes that under the current system, whistleblowers will receive only weak protection if they report corrupt activities.

### Strategy

The passage of Taiwan's whistleblower protection law in the third reading in the future will be the premise for whether this item receives a full score. To achieve a full score, it is necessary to clearly mention the complete protection of whistleblowers, including the protection of identity, prevention of retaliation, reversal of the burden of proof on retaliation, exemption of whistleblowers from liability, no sanctions for misleading reports, and the right of whistleblowers to refuse to participate in illegal acts.

Observing whether the defense department encourages whistleblowing is key to improving integrity in the future. From the full-score experience of Norway, although the Ministry of Defense also has its own reporting channels, which are not shared with other government departments, it is still necessary to provide evidence that there is an independent unit with sufficient resources to deal with reported appeal cases. It is necessary to hold a full-staff whistleblowing initiative that can cover personnel at all levels. It would be of substantial benefit if the Ministry of Defense could launch a comprehensive whistleblowing campaign in the near future, with a related whistleblowing procedure and protection training program.

Whether or not there is a high score on the effectiveness of whistleblower protection is not related to whether there is a whistleblower protection law in Switzerland; the focus is on whether there is evidence that the number of whistleblowing cases has increased year by year after the policy has been promoted. Only by having people report can whistleblowers' confidence that their reporting behavior is effectively protected be ensured.

## 4.2   Q37: Is Special Attention Paid to the Selection, Time in Post, and Oversight of Personnel in Sensitive Positions, Including Officials and Personnel in Defense Procurement, Contracting, Financial Management, and Commercial Management?

### Current Practice

In terms of the full coverage of high-risk positions, Taiwan currently only targets one or two areas that may be classified as high-risk targets (e.g., procurement) for personnel in sensitive positions, while other areas do not pay special attention. As far as the selection process is concerned, the provisions from the general selection process to the revolving door clause have been stated, but there is still a lack of a clear description of the selection process. From a supervisory point of view, the Ministry of Defence stated that internal oversight is responsible for reviewing appointment and promotion decisions of personnel in sensitive positions, as well as oversight and term limits for asset declarations by relevant sensitive personnel.

### Strategy

The scope of personnel in sensitive positions should be broadened, and in addition to procurement, positions with a high degree of discretion in recruitment, contracting, finance, and corporate management should be supported by relevant measures and included in special supervision. For example, in New Zealand, sensitive positions require higher security clearance, and thus, a higher threshold of supervision is necessary in the review process. Similarly, the Dutch approach involves special investigations into military intelligence and security personnel. The hiring process for senior positions involving procurement, appointment, contracting, finance, or business decisions is subject to not only general controls but also special hiring process controls.

This project is related to the practice in the preceding paragraph and further provides for the selection of personnel for high-agility positions, with more specific and objective provisions for the stages of interview, pre-selection, shortlisting, evaluation, and selection. Emphasis is placed on following a standard selection process for positions requiring specific technical competencies. Additionally, it is necessary to specify the number of years during which these astute officers are restricted from operations after leaving the Ministry of Defense.

From the experience of Switzerland, the feasibility and channels for anonymous reporting of corruption and illegal activities by vigilant personnel can be emphasized. This approach verifies the effectiveness of supervision of illegal activities from the perspective of vigilant personnel overseeing illegality. Another perspective is to learn from the practice of the Netherlands, which emphasizes that the review and appointment of high-level sensitive decision-making officials in defense should be conducted by an external objective committee of the Ministry of National Defense and then submitted to the Minister of Defense for decision-making.

### 4.3    Q38: Is the Number of Civilian and Military Personnel Accurately Known and Publicly Available?

**Current Practices**

In terms of the accuracy of the number of defense personnel, Taiwan is indirectly presumed to be available in the annual defense budget, that is, on the basis of an announcement once a year, so that only half of the points can be obtained. In terms of transparency, the Ministry of Defence currently publishes the number of military and civilian personnel in the form of annual aggregate or summary data, which also meets half of the transparency requirement of this item.

**Strategy**

This item focuses on the frequency of disclosure, because Taiwan still has the "Guidelines for the Classification of Military Secrets and National Defense Secrets", so in terms of the accuracy and transparency of the Ministry of National Defense's individual information announcements in the national armyThere are indeed legal limitations that cannot be overcome. However, the focus here is only on the immediacy of the information, so in the future, the way to moderately increase the score can still be in the form of aggregation, quarterly (4 points) or semi-annual (3 points) in the appropriate premises or websites. That is, in order to achieve full marks in the assessment of the accuracy of the number of personnel, the number of civilian and military personnel should be updated at least quarterly, and there should be established procedures for the publication and verification of statistical data. In the case of New Zealand, Germany and the United Kingdom, which provide the number of military civilian personnel in a quarterly or online format on a monthly basis, with statistical units to corroborate.

This item is aimed at whether to further provide information on the categories of civilian personnel of various militaries, which may be more limited by the norms of the "Guidelines for the Classification of the Scope and Classification of Military Secrets and National Defense Secrets" for Taiwan. However, whether further disclosure of transparency by category will violate the disclosure criteria should be consulted by the Legal Department of the Ministry of National Defence for discussion. The UK publishes information on military and civilian personnel by rank on a quarterly basis, while the number of civilian personnel in the German armed forces by rank is available on the website of the German armed forces to show that it has achieved full transparency of information category disclosure.

## 4.4   Q 41: Is There an Established, Independent, Transparent, and Objective Appointment System for the Selection of Military Personnel at Middle and Top Management Level?

*Current Practice*

Taiwan currently has objective standards and procedures in the formal selection process but does not have a mechanism for external observers or external participation in consultation. As far as the review mechanism is concerned, Taiwan's Legislative Yuan does not participate in the selection and deliberation of middle- and high-level military personnel, so this question can only score 1 point at most. In terms of the transparency of incumbent officers, in fact, most of Taiwan's middle- and high-ranking officers can be appointed according to the graduation of military academies and the length of appointment of each branch of the military to objectively estimate the relevant suitable candidates, and then compare the relative performance. Moreover, the appointment of middle- and high-ranking officers has been exposed in various newspapers and media.

*Strategy*

In the personnel procedures for selecting middle and high-ranking officers, Taiwan has a good system for objective selection, and it is not easy to obtain 3 points, and from the perspective of feasibility, the status quo should be maintained. In terms of the mechanism of external review, the selection and appointment of military personnel at any level in Taiwan does not need to be consulted by the Legislative Yuan, so it is recommended that the status quo should be maintained. In terms of transparency, Japan received a score of 1 in this evaluation, indicating that the evaluation team judged that the appointment information was almost not announced on the Internet. This is because from the point of view of the scoring criteria for this item, as long as the information about the selection criteria and the final appointment process for each middle and senior level is public, a full score will be awarded. In the case of the United Kingdom and Norway, which reveal information on the appointment process and formal selection criteria for each mid- to high-level post, which can be found on the website of the Norwegian Ministry of Defence, they received full marks.

## 4.5 Q 42: Are Personnel Promoted Through an Objective, Meritocratic Process? Such a Process Would Include Promotion Boards Outside of the Command Chain, Strong Formal Appraisal Processes, and Independent Oversight

### Current Practice

Almost all countries have low scores in this question, and only Switzerland has done the best and received full marks. Taiwan scored 0 points in this category, while Japan in the Asia–Pacific region performed well with 3 points. There are four sub-sections of this topic, namely (1) formal procedures, (2) exceptionality, (3) comprehensiveness, and (4) frequency. Taiwan has achieved full marks in (2) the exception of personnel promotions and (4) the frequency of personnel promotions, so they must be reviewed from the other two items. In (1) the formal procedure is scored as 3 points, that is, there are formal assessment requirements and procedures for the promotion of officers. Comprehensiveness received an incredible score of 0. Because the situation of receiving 0 points means that there is almost no information about the promotion of middle and high-ranking military and civilian officials!

### Strategy

In order to improve the practice of achieving full marks in formal procedures, it is necessary to set up external observers or external committees when promoting relevant defence officials, which is not currently possible for us, so it is recommended that the status quo be maintained. As for (3), the primary approach is to assume that this is based on the assumption of NATO countries, and by analogy whether countries publish information[1] on the promotion of officers at the OF-4 level and above. If Taiwan responds to Taiwan's failure to announce the promotion information of officers above the OF-4 level because it is not a member of NATO, it may be deemed that it has not released any information on the promotion of officers who have been exposed!

In the case of Switzerland and Japan, for example, promotions at the highest level in Switzerland are only reviewed by an external committee, but promotions from the rank of lieutenant are announced in a press release listing their names, new ranks, effective dates, place of residence, etc. In Japan, the promotion of colonel/general will be announced on the official website of the Ministry of National Defense, and the promotion information will not be announced at the same time as the corresponding level of civilian officials of the Ministry of Defense, but the latest list of civilian officials will be provided on the official website of the relevant Ministry of Defense.

The Ministry of National Defense of Taiwan will take the initiative and regularly announce the promotion of generals, which should be the primary focus of the response, so that the international evaluation units can understand that we must make

---

[1] The rank of OF-4 officer in NATO member states is approximately equivalent to the rank of lieutenant colonel in Taiwan.

a complete announcement (including the OF-4 level) for officers above the rank of general, and there are also channels and announcements for the promotion of other non-general and civilian officialsThe official personnel update on the website can be queried, and it should be possible to strive for a level of at least 2 points.

## 4.6   Q 45: Are Chains of Command Separate from Chains of Payment?

### Current Practice

Taiwan received 3 points on this question, and the reason for not getting the full score may be that the international assessors believe that the commander has substantive discretion in the part of the reward and the approval of the official account.

### Strategy

Looking at the practice of Germany, the United Kingdom and New Zealand, Germany cites the Federal Civil Service Remuneration Act ("Bundesbesoldungsordnung") to indicate that the chain of payment is indeed separated and allows military units to propose performance bonuses or allowances but must communicate with the appropriate human resources department/ Coordination. And the UK is more concise about the fact that the troop payment model is indeed separate from the chain of command, and the circumstantial source website is the source. In New Zealand, details are not only subject to specific laws (the Paycheck Protection Act), but also external audits by auditors. Therefore, Taiwan should be able to supplement the audit internal and external monitoring procedures to strive for a full score.

## 5   Conclusions

Corruption prevention has long been an international political priority, and various control mechanisms, such as the United Nations Convention against Corruption (UNCAC) and the Group of Anti-Corrupt Countries, have emphasized the need to initiate anti-corruption policies. Research shows that organizational-level analysis is increasingly used to investigate the determinants and possible measures of corruption prevention. The GDI's multi-year evaluation framework and comprehensive questions examine the state of defense corruption from a national perspective, and the results can be used by countries to learn from and compare with each other, thereby achieving the effect of preventing corruption risk. Therefore, conducting an in-depth self-examination based on the evaluation results of the GDI is of great significance for further improving integrity.

In the GDI 2020 evaluation item at the level of defense personnel risk, Taiwan received a Band A evaluation, which is a strength for Taiwan. Of the 17 topics on the risk of corruption in national defense personnel, only 6 failed to reach Band A, and only 3 of those failed to reach Band B. However, from the evaluation of GDI 2020 in Taiwan, I still specifically pointed out the potential risks of corruption in defense personnel, including the protection of whistleblowers in the Taiwan military, the verification of officials in important and agile positions, and the objectivity of the selection and promotion of senior officials in the Taiwan military. These are all areas that can be improved in the future. This chapter introduces the strongest aspects of Taiwan's GDI, hoping to serve as a model for other countries. However, among the five major aspects evaluated, the military operation integrity score only received a Band D evaluation, which is still far behind New Zealand's Band A evaluation. This is the most important part that deserves in-depth study in the future. By following the methods outlined in this chapter to improve the various items of the GDI, I believe there will be better results in the GDI's 2025 integrity assessment. Most importantly, in the nature of anti-corruption, there will be more pragmatic self-improvement results.

# References

Beeri I, Dayan R, Vigoda-Gadot E, Werner SB (2013) Advancing ethics in public organizations: the impact of an ethics program on employees' perceptions and behaviors in a regional council. J Bus Ethics 112(1):59–78

Berman EM (2015) HRM in development: lessons and frontiers. Publ Administr Dev 35(2):113–127

Burlando A, Motta A (2016) Legalize, tax, and deter: optimal enforcement policies for corruptible officials. J Dev Econ 118:207–215

Campbell JW (2020) Buying the honor of thieves? Performance pay, political patronage, and corruption. Int J Law Crime Justice 63:100439

Capasso S, Goel RK, Saunoris JW (2019) Is it the gums, teeth or the bite? Effectiveness of dimensions of enforcement in curbing corruption. Econ Govern 20(4):329–369

Ceschel F, Hinna A, Homberg F (2022) Public sector strategies in curbing corruption: a review of the literature. Publ Organiz Rev 22:571–591. https://doi.org/10.1007/s11115-022-00639-4

Chang Y, Wilding M, Shin MC (2017) Determinants of whistleblowing intention: evidence from the South Korean government. Publ Perform Manag Rev 40(4):676–700. https://doi.org/10.1080/15309576.2017.1318761

Chen Y, Liu Q (2018) Public-sector wages and corruption: an empirical study. Eur J Polit Econ 54:189–197

Devine T, Maassarani T (2011) The corporate whistleblower's survival guide: a handbook for committing the truth. Berrett-Koehler, San Francisco

Dimant E, Tosato G (2018) Causes and effects of corruption: what has past decade's empirical research taught us? A Survey. J Econ Surv 32(2):335–356

European Parliament (2017) Resolution of 24 October 2017 on legitimate measures to protect whistle-blowers acting in the public interest when disclosing the confidential information of companies and public bodies (2016/2224(INI)). https://www.europarl.europa.eu/doceo/document/TA-8-2017-0402_EN.html. Accessed 2 Mar 2023

Fišar M, Krčál O, Staněk R, Špalek J (2021) Committed to reciprocate on a bribe or blow the whistle: the effects of periodical staff-rotation in public administration. Public Perform Manag Rev 44(2):404–424

Grant A (2021) Think again: the power of knowing what you don't know. Penguin

Graycar A, Monaghan O (2015) Rich country corruption. Int J Public Adm 38(8):586–595

Haywood LK, Funke N, Audouin M, Musvoto C, Nahman A (2019) The sustainable development goals in South Africa: investigating the need for multi-stakeholder partnerships. Dev South Afr 36(5):555–569

Holt D, Davis S (2022) Interrupting bias in army talent management. US Army War College Q Parameters 52(1):21–40

Horton A (2020) Veterans with 'bad paper' turned away from VA for decades, Harvard Law study finds. The Washington Post

IBA (2021) Are whistleblowing laws working? A global study of whistleblower protection litigation. International Bar Association, London

Janenova S, Knox C (2020) Combatting corruption in Kazakhstan: a role for ethics commissioners? Publ Administr Dev 40(3):186–195

Joy M (2019) Powerarchy: understanding the psychology of oppression for social transformation. Berrett-Koehler Publishers

Klitgaard RE, Abaroa RM, Parris HL (2000) Corrupt cities: a practical guide to cure and prevention. World Bank Publications

Luk SCY (2012) Questions of ethics in public sector management: the case study of Hong Kong. Publ Personnel Manage 41(2):361–378

Mangafić J, Veselinović L (2020) The determinants of corruption at the individual level: evidence from Bosnia-Herzegovina. Econ Res-Ekonomska Istraživanja 33(1):2670–2691

Munro I, Kenny K (2023) Networked whistleblowing, counter-hegemony and the challenge to systemic corruption. In: Organizational wrongdoing as the "foundational" grand challenge: consequences and impact, vol 85. Emerald Publishing Limited, pp 121–140

Navot D, Reingewertz Y, Cohen N (2016) Speed or greed? High wages and corruption among public servants. Administr Soc 48(5):580–601

OECD (2019) ANNEX: G20 compendium of best practices and guiding principles for legislation on the protection of whistleblowers. In: G20 anti-corruption action plan protection of whistleblowers. https://www.oecd.org/g20/topics/anti-corruption/48972967.pdf. Accessed 28 May 2023

Olsen AL, Hjorth F, Harmon N, Barfort S (2019) Behavioral dishonesty in the public sector. J Publ Administr Res Theor 29(4):572–590

Pyakuryal S, Cox RW (2019) Corruption in South Asia. In: Corruption in a global context. Routledge, pp 126–139

Pyman M (2023) Tackling defense corruption. Ethical dilemmas in the global defense industry, 259

Sylvester R (2017) Don't shoot the messenger. Whistleblowers and the all-too-human... By Ray Sylvester. Hyperlink Magazine. Medium. Retrieved 18 June 2023, from https://medium.com/hyperlink-mag/dont-shoot-the-messenger-384ebad1369b

Thaler J, Helmig B (2016) Do codes of conduct and ethical leadership influence public employees' attitudes and behaviours? An experimental analysis. Publ Manage Rev 18(9):1365–1399

Transparency International Ireland (2017) Speak up report. https://transparency.ie/resources/whistleblowing/speak-report-2017

Transparency International UK Defense, & Security Program (TI-DS) (2021) Government defense integrity index 2020 (GDI 2020). UK

Tsao YC (2023) Reduce the risk of corruption in military operations. Publ Affairs Rev 20(1):1–17

Tsao YC, Hsueh SJ (2023) Can the country's perception of corruption change? Evidence of corruption perception index. Publ Integr 25(4):415–427

US Army Talent Management (2020) Commander's guide to ATAP. Department of the Army, Washington, DC, p 13

Van Veldhuizen R (2013) The influence of wages on public officials' corruptibility: a laboratory investigation. J Econ Psychol 39:341–356

Wharton (2019) Whistleblowers in business: do the risks outweigh the benefits? Knowl-edge@Wharton. Retrieved 1 Sept 2023, from https://knowledge.wharton.upenn.edu/article/whi stleblowers-in-business/

Więcek-Starczyńska A, Mroczek-Dąbrowska K, Trąpczyński P (2019) Multidimensional compar-ative analysis of corruption in Europe from 1999–2013. In Corruption in a Global Context (pp. 201–220). Routledge

# An In-Depth Analysis of Morocco's New Strategies for Agricultural Transformation and Industrializing Agriculture for Sustainability and Peace

Oumaima Tounchibine⑩, Hindou Baddih, Fatima Zohra Azizi, and Fouzia Majidi

**Abstract** Morocco has invested heavily in agricultural innovations to enhance productivity and yield. However, the sector faces challenges of water scarcity, climate change, and smallholder farmers. Industrializing agriculture through advanced irrigation systems, manufacturing plants, and access to resources for smallholders can address these challenges. This study examines how industrializing Moroccan agriculture can promote sustainability and peace. First, it can improve economic and environmental sustainability by increasing the sector's overall impact and diversifying the economy. Second, it can generate new jobs, reducing unemployment that is a source of social unrest. Third, industrialization can help smallholders cope with challenges through access to technology, financing, and markets, improving their livelihoods and wellbeing. The findings suggest that industrializing agriculture, if done equitably and sustainably, can contribute to both economic development and social peace in Morocco by making the sector more resilient and inclusive. However, careful planning, governance reforms, and investment in smallholders will

---

O. Tounchibine (✉) · H. Baddih
Laboratory of Economic Sciences and Public Policy, Department of Economics and Management Sciences, Faculty of Economics and Management, Ibn Tofail University, Kenitra, Morocco
e-mail: t.oumaima@live.iium.edu.my

H. Baddih
e-mail: Baddih.hindou@uit.ac.ma

*Present Address:*
O. Tounchibine
Institute of Islamic Banking and Finance, International Islamic University Malaysia (IIUM), Kuala Lumpur, Malaysia

F. Z. Azizi
Mohammed V University, University Institute of African, Euro-Mediterranean, and Ibero-American Studies, Rabat, Morocco
e-mail: fatima-zohra.azizi@um5.ac.ma

F. Majidi
Department of Legal, Economic, and Social Sciences, Mohammed 1 University, Oujda, Morocco
e-mail: fouzia.majidi@ump.ac.ma

217
A. Sharifi et al. (eds.), *Navigating Peace and Sustainability in an Increasingly Complex World*, World Sustainability Series, https://doi.org/10.1007/978-981-97-8772-2_11

be needed to ensure the benefits of industrialization are widely shared. The study contributes to the literature on agricultural transformation and development by highlighting how industrialization strategies can promote sustainability and social goals, not just economic growth. It also provides practical recommendations for Moroccan policymakers seeking to advance these dual objectives through agricultural policy.

**Keywords** Agriculture · Morocco · Agriculture economics · Innovation · Industrialization

## 1  Introduction

Morocco has long been aware of the historical heritage of its agriculture, which has seen a spectacular expansion in recent years. Today, it is one of the main food producers and an economic driver in North Africa (Ghanem 2015). Due to its favorable geographical location, including coastal plains, fertile valleys, mountainous regions, and desert areas, Morocco has a wide variety of agricultural products. Additionally, the country benefits from the Mediterranean Sea, a vast coastline, and the Atlas Mountains, which positively influence agricultural production systems and their diversity (Montanari 2012).

Currently, Morocco utilizes 30 million hectares of land for agriculture. Seven million hectares are arable land, while over 21 million hectares are covered by permanent pastures and meadows. The remaining 1.7 million hectares are dedicated to permanent crops (Statista Research Department 2024b). In addition to agricultural production, Morocco is also engaged in significant agricultural and fisheries operations as a nation. Moroccan agriculture consists of three main sectors: (i) contemporary, private, and irrigated agriculture focused on exports and the cultivation of mainly fruits and vegetables; (ii) production of dairy products, sugar, seeds, fruits, and vegetables in large irrigated agricultural areas by dams, mainly for the local market; and (iii) rain-fed agriculture in the northwest for the production of cereals, olives, legumes, red meat, and dairy products, and in the south and east for the same crops, mainly cereals, and non-intensive sheep production (Office of Agricultural Affairs 2024).

Despite the advantages of Morocco's agricultural resources, the country still imports 25% of its cereals, primarily wheat, from Ukraine and Russia in 2023 (Erraji-Morocco World News 2024). Additionally, the majority of farmers continue to rely on traditional methods and struggle to access production inputs such as fertilizers, herbicides, and machinery. Moreover, small-scale farms are predominant, and the complexity of land ownership poses challenges, while rising land prices create major obstacles for stakeholders in the agricultural sector.

It is important to note that in recent years, the contribution of this sector has begun to decline due to two major events: the global COVID-19 pandemic (World Bank Group 2022) and the recent armed conflict between Russia and Ukraine (Barry et al. 2023). These invasions have negative implications. Firstly, they disrupt the supply of cereals and vegetable oils, mainly produced in Russia and Ukraine. Secondly, due to the halt or fear of an energy supply shortage from Russia, prices of coal, oil, and gas increase, leading to a direct increase in low-income households' expenses for heating, cooking, transportation, etc. Cascade economic effects are clearly observable in Morocco. The increase in energy source prices has a direct or indirect impact on food prices, leading to higher production and transportation costs in the agricultural sector, as well as higher costs for energy-intensive fertilizers. Some energy expenses cannot be reduced without causing serious difficulties. These two crises have had a negative impact on several sectors in Morocco, especially the agricultural sector, leading to an increase in inflation and exacerbating the situation (Ait Ali et al. 2022).

Hence, it is essential for the Moroccan government and all stakeholders to take innovative initiatives and measures to promote agriculture in Morocco and increase its production efficiency to become independent from other countries. There is a need for the modernization of the sector and the integration of innovations and new technologies to enhance its competitiveness and profitability. To achieve this purpose, the Moroccan government has launched several projects and programs (The Agency for Agricultural Development (ADA) 2020). They are aimed at addressing current challenges in the development of the sector to become independent of international vulnerabilities and promote peace and sustainability in its agricultural sector by focusing on (i) modernization and technological integration, (ii) enhancing self-sufficiency, and (iii) building resilience.

In this chapter, the researchers base the analysis on the theory of endogenous growth. This hypothesis suggests that an internal variable, indicating structural transformation, can contribute to understanding the endogenous dynamics of the agricultural sector and strengthen its performance in order to face the negative impact of external factors such as the global health crisis, drought, and the Russia-Ukraine war. Thus, by using a meta-analysis method, this chapter aims to examine the outcomes of the projects and programs intended to promote factors of peace and sustainability in the agriculture sector in Morocco. This will help define strategic directions that assist the Moroccan government in overcoming the mentioned challenges, with a focus on robust policies, integrating innovations and new technologies, and enhancing the sector's competitiveness to spread peace and sustainability. By addressing these areas, the Moroccan government and stakeholders can promote the industrialization of the agricultural sector in a way that enhances the country's peace, self-sufficiency, and sustainability in the long term.

## 2 Morocco's Economic Transformation During Drought

The contribution of agriculture in Morocco has shown significant variation from 2019 to 2023. Despite experiencing a general decline between 2019 and 2020, the sector saw a peak contribution in early 2021. By the end of 2023, the agriculture, forestry, and fishing sectors collectively contributed approximately 12% to Morocco's economy. This highlights the sector's vital role and resilience in the national economy, despite fluctuations in its performance over the years (Statista Research Department 2024a). The agriculture sector provides employment to 30% of the population (Rahhou 2023), with nearly 74% of jobs in rural areas being in agriculture (Oxford Business Group 2020). Recently, Morocco has confronted an extreme drought, which has coincided with the Russian-Ukrainian conflict and the continuous COVID-19 health crisis. The environmental change adversely affects the nation, prompting droughts that have caused significant damage and losses in the Moroccan agriculture sector (Ayham 2022).

According to projections from Morocco's 4th National Communication, the country will face significant declines in average rainfall and precipitation by 2050 (UNDP 2017). The majority of Morocco's national territory is expected to experience a decrease in rainfall and precipitation (Bouramdane 2024). This outlook of hotter and drier conditions will exacerbate the country's vulnerability to more severe and recurring droughts. Drought events that damage crops and livestock can have significant economic and social impacts in a country where agriculture plays such a vital role. Rural populations dependent on farming are disproportionately affected by agricultural losses from drought. Therefore, Morocco is committed to building resilience to droughts and managing resulting loss and damage in its agriculturally-dependent and climate-vulnerable rural areas.

In response to this threat, the United Nations Framework Convention on Climate Change (UNFCCC) coordinated a Transitional Committee workshop on loss and damage on April 29, 2023 (UNFCCC 2023b). The workshop aimed to fortify the worldwide response to environmental change and address the related damage and losses generated by this drought. During the workshop, the country stated that the increase in the frequency and intensity of drought events across the nation could be mainly attributed to climate change caused by greenhouse gas emissions from human activities. The situation in the country has been worsening because of human-caused climate change. The national context highlights the importance of Morocco's efforts to build resilience and manage loss and damage associated with worsening drought conditions driven by climate change (Moussaid et al. 2023a, b).

Morocco's Ministry of Agriculture, Maritime Fisheries, Rural Development, and Water and Forests reported a 67% decrease in cereal production for the 2021–2022 agricultural year due to drought and water stress. As a result, the value added from agriculture contracted by approximately 15%, and Morocco's reliance on imported cereals increased. These losses highlight the country's continued vulnerability to droughts and water scarcity, which can severely damage crop yields, even with agricultural development plans in place. The situation underscores the challenges that Morocco faces in building resilience and managing loss and damage caused by worsening drought conditions. The economic impacts reported provide valuable insights for other countries seeking to understand and minimize the agricultural losses that droughts can cause, despite preparedness efforts.

The World Bank has also observed that the current rainfall deficits are causing job losses in Morocco's rural areas (The World Bank 2022). According to data from Morocco's High Commission on Planning (HCP), 174,000 jobs were lost in rural areas in 2022, coinciding with the worst drought in 40 years (HCP 2023). These employment losses underscore the vulnerability of Morocco's rural labor market and livelihoods to drought conditions, which severely impact agricultural outputs. It highlights the challenges that Morocco faces in safeguarding rural jobs and incomes from worsening droughts, despite its efforts in adaptation and development. Overall, the reported job losses provide valuable insights for other countries seeking to understand and mitigate the impacts of drought on employment and rural economic development.

## 3 Morocco's Economic Transformation During Drought Period: Analyzing the Dynamic Agricultural Sector and Macro Poverty Indicators (2018–2023)

Between 2018 and 2023, Morocco experienced significant economic changes influenced by persistent drought conditions. This period highlighted the resilience and challenges faced by various sectors, particularly agriculture. By analyzing macro poverty indicators and sectoral growth rates, this study provides insights into the broader economic trends and the impact of drought on Morocco's economy. The data reveals fluctuations in real GDP growth, with notable contractions during drought years and periods of recovery. The agricultural sector showed dramatic variations in growth, while the industrial and services sectors exhibited relative stability and resilience. Through this comprehensive analysis, we aim to understand the dynamic interplay between agricultural performance and overall economic health, particularly in the face of environmental adversities.

### 3.1 Real GDP Growth

This is the measure of a country's economic performance by representing the percentage change in the country's Gross Domestic Product (GDP) after the removal of the effects of inflation. In Table 1, this term reflects the adjusted economic growth rate, providing a more accurate representation of the economy's health over time.

**Table 1** Macro poverty outlook indicators

| Indicators | 2018 | 2019 | 2020 | 2021 | 2022 | 2023 |
|---|---|---|---|---|---|---|
| Real GDP growth, at constant factor prices | 3.0 | 1.8 | − 6.1 | 5.4 | 3.3 | 3.5 |
| Agriculture | 2.4 | − 4.6 | − 6.9 | 17.6 | − 2.0 | 3.1 |
| Industry | 3.0 | 3.6 | − 3.8 | 3.1 | 3.1 | 3.5 |
| Services | 3.1 | 2.7 | − 6.8 | 3.3 | 4.9 | 3.7 |

World Bank Group, 2023

In 2018, the real GDP growth rate stood at 3.0%, indicating a robust and healthy pace of economic expansion. However, a noticeable slowdown occurred in 2019, with the growth rate dropping to 1.8% compared to previous years. This downward trend continued in 2020, when the economy experienced a significant contraction of − 6.1%, suggesting that the economy entered a recession. This recession was likely due to various factors, including the impact of the COVID-19 pandemic.

Fortunately, the economy bounced back in 2021 with a strong growth rate of 5.4%, indicating a recovery from the previous year's downturn. In the following years, 2022 and 2023, the growth rates are projected to be 3.3% and 3.5% respectively, indicating a more modest but still positive growth trend.

Overall, Table 1 shows that the economy experienced a decrease in the growth rate, followed by a recession and subsequent recovery. We expect continued but slower growth in the coming years.

### 3.2 Agricultural Growth in Morocco

Agriculture is a significant sector in many economies, and its performance can have a direct impact on food security and employment. In 2018, agriculture in Morocco showed a growth rate of 2.4%, which is positive. However, it experienced a severe contraction of − 4.6% in 2019, followed by an even more significant decline of − 6.9% in 2020. The sector then rebounded strongly in 2021 with a growth rate of 17.6%, possibly due to factors like favorable weather conditions or government interventions.

In 2022, agriculture saw a minor decline of − 2.0%, but it is expected to grow at a more sustainable rate of 3.1% in 2023. Overall, agriculture growth in Morocco exhibits significant volatility, with fluctuations ranging from a contraction of − 6.9% to an expansion of 17.6%. This variability is indicative of the sector's high dependence on external factors, primarily weather conditions and commodity prices, which can significantly influence agricultural outcomes.

### 3.3 Industry Growth in Morocco

The industrial sector, which includes manufacturing, mining, and construction, is another critical component of the economy. In 2018, it grew by 3.0% and further expanded by 3.6% in 2019, indicating robust economic growth. However, in 2020, the sector contracted by − 3.8%, reflecting the economic challenges during that year. This contraction is primarily attributed to the adverse impacts of the COVID-19 pandemic, which disrupted industrial production and construction activities.

The industry recovered in 2021 with a growth rate of 3.1%, and it is projected to maintain a growth rate of 3.5% in 2022 and 2023, suggesting a more stable and positive outlook. This indicates that the industrial sector in Morocco is on a path of stability and steady expansion during these years.

Overall, the data reveals a consistent and resilient industrial sector that, despite facing challenges in 2020, is expected to recover and maintain stable growth, contributing to Morocco's economic development.

### 3.4  Service Sector Growth in Morocco

The services sector encompasses a wide range of activities, including finance, retail, healthcare, and more. In 2018, the services sector grew at a rate of 3.1%. It experienced slower growth of 2.7% in 2019 and a substantial contraction of $-6.8\%$ in 2020, likely influenced by the pandemic's effects on travel and hospitality. However, services rebounded in 2021 with a growth rate of 3.3%. The sector is expected to grow at a healthy pace in 2022 and 2023, with projected growth rates of 4.9% and 3.7%, respectively.

While the service sector, like other sectors, faced a setback in 2020 due to the pandemic, it is poised to recover and continue its expansion, contributing to Morocco's economic development and diversity.

In summary, Table 1 provides an overview of the economic performance in various sectors over the specified years. It shows the impact of external factors, such as the global pandemic, on the economy. The data suggests a mixed picture with periods of growth and contraction, but the outlook for 2022 and 2023 appears to be more stable and positive, with most sectors showing positive growth rates.

## 4  Main Outputs of Different Moroccan Ministries Regarding Loss and Damage Related to Drought

Different ministries in Morocco shared their experiences regarding the management and financial aspects of dealing with the losses and damages caused by the drought. In this section, we will provide insights into the current situation in Morocco, how the government reacts to and deals with the situation, and the measures taken by the Moroccan government to reduce the negative impact of this issue, as presented by the ministries of Morocco to the UNFCCC.

### 4.1 Morocco's Ministry of Energy Transition and Sustainable Development: Managing Loss and Damage Related to Drought

The UNFCCC reported that the Ministry of Energy Transition and Sustainable Development of Morocco shared its valuable experiences and insights regarding losses and damages associated with drought. The ministry emphasized its commitment to energy transition and sustainable development as key strategies for enhancing resilience and addressing losses and damages resulting from climate impacts. In the realm of energy transition, Morocco has outlined significant renewable energy programs scheduled between 2016 and 2030. These projects aim to accomplish the addition of 10,100 MW of renewable energy capacity, incorporating solar (20%), wind (20%), and hydro (10%) (UNFCCC 2023a).

In the context of sustainable development, Morocco is effectively laying the foundation for a green and inclusive economy by the year 2030 (Chafil 2018). This initiative involves the implementation of its Nationally Determined Contribution under the Paris Agreement and its National Adaptation Plan. In summary, the workshop offered Morocco a valuable platform to share the lessons it has learned from its policies and strategies for bolstering climate resilience and effectively managing losses and damages, particularly in the face of recurring drought events. The experiences and insights shared by Morocco can serve as a valuable resource for other nations aiming to strengthen their capabilities in the management and financing of loss and damage.

### 4.2 Morocco's Ministry of Agriculture, Marine Fisheries, Rural Development, and Water and Forestry: Strategies for Agricultural and Water Management

In the workshop held by UNFCCC, the Ministry of Agriculture, Marine Fisheries, Rural Development, Water, and Forestry of Morocco presented the country's significant agriculture and water-related plans and initiatives. It provided a general overview of the Plan Maroc Vert from 2008 to 2020. The plan aimed to achieve two primary goals: promoting modern and high-value agriculture that adheres to international standards, and assisting small-scale agriculture to alleviate rural poverty and foster income diversification.

The ministry also presented its Green Generation 2020–2030 plan, which depends on two core principles. The first principle is to focus on the well-being of individuals and place them at the forefront of concerns. The second principle is to promote the long-term sustainability of agricultural development (UNFCCC 2023a).

Morocco is actively working to enhance resilience in the agriculture and water sectors through initiatives such as Plan Maroc Vert and Green Generation. By managing loss and damage from climate impacts like drought, the country is taking steps to ensure food security and rural development. Other nations can gain valuable insights from Morocco's experiences in implementing effective agriculture and water management plans amidst the challenges of climate change.

## 4.3 Morocco's Ministry of Equipment and Water: Infrastructure and Water Management Initiatives

The Ministry of Equipment and Water of Morocco discussed the country's water management plans and strategies in the Transitional Committee workshop on loss and damage held by UNFCCC. Morocco emphasized key water management priorities such as constructing dams, interconnecting river basins, and desalinating seawater to improve water supply. Other priorities include integrating all rural centers into structured drinking water supply systems, providing necessary water resources to enable sustainable agricultural development, preserving ecosystems, and fighting pollution (ITA 2022).

Morocco also presented its National Water Plan 2020–2050, which aims to strengthen integrated and sustainable water resource management (UNFCCC 2023a). Through strategies like this plan, Morocco is working to improve climate resilience and manage loss and damage related to water scarcity and drought. The country's experiences in managing water resources in the face of climate change can inform policies and approaches in other nations facing water security challenges.

Overall, infrastructure to improve water storage and irrigation is critical to coping with water scarcity. However, international experience suggests that "engineering solutions" need to be combined with effective policies to manage water demand.

## 5 Strategic Measures: The Moroccan Government's Initiatives to Tackle Drought Impacts in the Agriculture Sector

Morocco has implemented various measures to manage the losses and damages caused by the severe 2021–22 drought. The country's response highlights potential channels for deploying international climate finance support. Morocco's use of public budget allocations for the PRIDP program emphasizes the importance of national social protection systems and safety nets that can be scaled up with international support in the face of climate change risks. The creation of an emergency relief fund through PRIDP also demonstrates how dedicated emergency funds drawing on national and international resources can provide rapid response assistance after climate disasters. Furthermore, Morocco's support for agricultural insurance shows the potential for international finance to subsidize and strengthen national catastrophe risk insurance systems covering climate hazards like drought (UNFCCC 2023a).

The experience of Morocco reveals how international climate finance directed through national systems like social protection, emergency funds, and insurance can potentially help countries manage escalating climate change loss and damage. These national channels could inform the development of the Santiago Network for Loss and Damage as it works to catalyze technical assistance and finance.

To address the unprecedented drought, the Ministry of Agriculture in Morocco launched the Integrated Program for Drought Impact Mitigation (PRIDP), allocating a budget of 10 billion Moroccan dirhams (approximately $1 billion USD). The primary objective of the PRIDP is to minimize the impact of the drought on the agricultural season of 2021–2022 by providing support to farmers and breeders in three key areas: (1) Agricultural insurance and financial assistance for farmers and agriculture professionals, (2) effective management of water scarcity, and (3) protection of animal and plant assets.

Morocco has reported the disbursement of approximately $1 billion USD in public funds through the PRIDP to address the losses and damages caused by the recent severe drought. The establishment and funding of the PRIDP highlight Morocco's proactive approach in managing and responding to agricultural damages, job losses, and the economic repercussions of worsening drought conditions. This initiative provides valuable insights into policy and financial measures that other countries can consider to tackle loss and damage effectively (UNFCCC 2023a).

Morocco supports the establishment of a loss and damage fund at COP27 due to significant drought-related losses and damages that have impacted its agricultural yields, financial value, labor market, and livelihoods (WBG n.d.). The UNFCCC loss and damage fund presents an opportunity to reinforce Morocco's current national financing measures that have been put in place to address escalating loss and damage. However, it is crucial to conduct further research and provide support to improve existing national financing schemes that rely on domestic public resources in developing countries like Morocco. As climate change continues to harm vulnerable nations, reliable and adequate financing provided through effective national instruments can help ensure a fair response.

## 6 Empirical Evidence and Outcomes of Moroccan Efforts and Projects to Develop Agricultural Industrialization for Peace and Sustainability

The Moroccan government has implemented various initiatives to increase the profitability of its agricultural sector and to have significant economic impacts on the country. Several studies have highlighted Morocco's efforts to make the agricultural sector more attractive, both domestically and internationally (Adghough 2018; Asedrem 2021; Kydd and Thoyer 1992; Mattia 2021; Michel-Kerjan et al. 2014).

According to a report published by Tomay at the OECD, the Moroccan government has launched the Green Morocco Plan, also known as the Plan Maroc Vert, with the aim of promoting the agricultural sector and using it as a catalyst for the country's economic growth. To achieve this goal, the government has implemented various initiatives described in the plan to facilitate the expansion of the sector. The plan focuses on improving the regulatory framework in various aspects related to agriculture, including legislative, institutional, and regulatory contexts. This is aimed at enhancing the sector's appeal and attracting potential international investors. The plan also aims to improve farmers' access to markets and financial resources, as well as their competitiveness and productivity. It encourages the sustainable management of the agricultural environment. In summary, the Plan Maroc Vert is a comprehensive strategic framework aimed at transforming the Moroccan agricultural sector into a key driver of economic growth and development in Morocco. By implementing these initiatives, the government hopes to promote a more sustainable and prosperous future for the country (Mattia 2021).

Another recent study conducted in 2021 by Asedrem highlighted the efforts made by the Moroccan government to improve the responsiveness and productivity of the agricultural sector. The research detailed the various strategic objectives targeted by the Plan Maroc Vert implemented by the government to boost the sector. These strategic planning objectives encompass areas related to agricultural modernization, socio-economic development, sectoral framework redesign, natural resource preservation (including water conservation), adoption of clean energy and technologies, biodiversity, resilience to climate change, soil fertility, and land capacity mapping. To implement these strategic planning objectives and support them, the government has collected data on land capacity and soil fertility and has implemented various modern and ecological technologies, such as the establishment of water and soil conservation techniques, irrigation systems, improvements to abattoirs, dissemination of certified seeds, fertilization of agricultural land, as well as projects focused on crop intensification and conversion (Asedrem 2021).

The most crucial strategic objective of the plan is to improve the conditions of small-scale farmers through participatory and contractual approaches. This includes initiatives aimed at promoting balanced relationships between small-scale farmers, financial markets, and price stability; diversifying agricultural production through skills and technology transfer; and increasing animal and plant production. It introduces three essential concepts in the field of agriculture: the agricultural aggregation project (AAP), solidarity support agriculture, and land improvement for pastoral use. The agricultural aggregation strategy is a crucial tool aimed at bringing farmers together to implement agricultural investment projects in order to cultivate collaborations with various stakeholders from the private and public sectors within the agricultural industry (El Houssain et al. 2019). This is based on the concept of establishing contractual partnerships between small and medium-sized agricultural businesses and large-scale farms to create large-scale agricultural clusters. The Moroccan policy aims to expand the scale of agricultural businesses to enable better access to financial resources for investing in.

The main goal of the Moroccan government is to adapt legal frameworks to stimulate agricultural growth and prioritize its advancement (Arndt and Tyner 2003). It has updated and strengthened its legal frameworks to meet the changing needs of the agricultural strategy in order to modernize the sector, make it more productive and attractive, and ensure the sustainable development of rural areas (Abdelmajid et al. 2021). It is important to highlight the recent substantial efforts of the Moroccan ministry and its significant role in the agricultural sector to indicate the effectiveness of the ministry's agricultural policies in recent years. In line with the conclusions reported by the African Development Bank Group, the bank plays a significant role in shaping Moroccan agricultural policy (Bertin et al. 2021). Crucial achievements have been made through the implementation of four major agricultural programs from 2009 to 2019. These programs are primarily the Support Program for the Green Morocco Plan (PAPMV), the National Program for Irrigation Water Conservation (PAPNEEI), Inclusive and Sustainable Development Support Program for Agricultural Sectors (PADIDFA), and the Project for the Preservation and Development of Socio-Territorial Oases in the South (ADBG n.d.). Overall, the report represents the main purpose of each program, highlighting the results and benefits of these programs for Morocco's economic agriculture. It shows that the Moroccan government is updating its policies in favor of the agricultural sector to encourage its development in order to contribute to the country's sustainability. In summary, the support programs generate multiple on-the-ground benefits, including better water management, improved agricultural practices, increased employment opportunities, women's empowerment in agriculture, value chain development, and improved accessibility and transportation. Overall, the African Development Bank (AfDB) plays a central role as one of the institutions actively involved in providing assistance and support for this strategic agricultural initiative.

On November 20, 2020, the International Bank for Reconstruction and Development (IBRD), a part of the World Bank Group, released a program evaluation document describing a proposed $250 million loan to the Kingdom of Morocco to support Morocco's Green Generation for Results Program (P170419) (IBRD and WBG 2020). The Ministry of Agriculture, Fisheries, Rural Development, Water, and Forests was responsible for program implementation. This program was launched based on the Green Generation Strategy (GGV) 2020–2030, initiated by King Mohammed VI in February 2020 in response to the COVID-19 crisis, aiming to support the development of the Moroccan agricultural sector. The proposed Green Generation for Results (PforR) Program was co-financed by the World Bank and the Agence Française de Développement (AFD), with a project approval target date set for December 15, 2020, and a projected closing date of December 31, 2025 (Diaz Cassou et al. 2022). The strategy is designed to address critical issues related to immediate and long-term economic challenges due to the impact of COVID-19 on the country's economy and aims to prevent and reduce difficulties caused by drought while supporting affected communities. The main objectives of this strategy include promoting economic integration of youth in rural areas and enhancing marketing efficiency and environmental sustainability of agri-food value chains in a specific area. In summary, King Mohammed VI launched the GGV as a people-centered approach aiming to:

- Adopt a different perspective by shifting away from a production-focused strategy to focus on people, with the aim of stimulating economic growth and creating employment opportunities in Morocco's rural areas.
- Strongly emphasize expanding the middle class in rural regions, creating job opportunities for youth in both the agricultural sector and beyond, and promoting human capital development and skill acquisition to improve higher-paying jobs and work flexibility.
- Set the goal of accelerating the transition to a more climate-resilient agriculture, as initially envisioned in the Green Morocco Plan (PMV).
- Comprehensively address crucial actions needed to address the urgent COVID-19 pandemic and enduring structural obstacles faced by agriculture, such as recurring droughts.
- Provide a vital path for job creation and financial progress in disadvantaged rural regions of Morocco, crucial for national recovery after the pandemic.
- Prioritize the growth of human capital and skills, and increase labor productivity and mobility, which can ultimately contribute to higher income levels and a better standard of living for rural inhabitants.

In general, this policy is considered a flexible extension and a new strategic approach to the Green Morocco Plan (PMV), aiming to strengthen and expand its beneficial influence on the agricultural sector. It marks an important step towards achieving the sustainable development goals (SDGs) defined in international agreements such as the UN's 2030 Agenda for Sustainable Development and the Paris Agreement on climate change.

A recent report published by the World Bank Group in 2023 examined the World Bank Group's commitment to Morocco during the fiscal years 2011–2021. The World Bank used monitoring and evaluation practices to adjust the focus of Irrigated Agriculture Modernization through various projects. This led to a shift from infrastructure development in 2012 to water management and modern irrigation technologies in 2014, ultimately incorporating climate adaptation techniques in 2021. The Bank's adaptive approach ensured that the series was aligned with the changing needs and priorities of the Moroccan agricultural sector (World Bank 2023).

Therefore, it is important to emphasize that the World Bank has adopted a flexible approach to maintain long-term engagement with the government in various thematic areas, including agricultural policy. This plays a crucial role in monitoring and evaluating the Moroccan agricultural sector.

It is undeniably evident that the Moroccan government has shown remarkable dedication to improving the development of the agricultural sector through the diligent implementation of two prominent agricultural policies and strategies: the Green Morocco Plan 2008–2020 and the Green Generation Strategies 2020–2030. These initiatives undeniably demonstrate the government's steadfast commitment to promoting the growth and advancement of the agricultural sector. The indicated strategies and their evolution over time are discussed to provide a comprehensive understanding of government initiatives. Although these ongoing strategies represent the main policies and initiatives undertaken by the Moroccan government, it is recognized that there are many additional details and subsequent strategies that have been initiated based on these two policies. These additional strategies aim to translate the main objectives into concrete actions, thereby promoting the achievement of the Sustainable Development Goals. Therefore, it is crucial to conduct a systematic review to obtain a comprehensive understanding of these additional details and subsequent strategies in order to identify potential gaps that can serve as a basis for future studies.

The United Nations High-Level Political Forum on Sustainable Development (HLPF) published a comprehensive national report in 2020 on its official website. Titled "Voluntary National Review of the Implementation of the Sustainable Development Goals," this report was produced by the High Commission for Planning of Morocco in response to the UN's call for countries to assess their progress in achieving the SDGs. This call follows the review and revision of the 2030 Agenda for Sustainable Development and the SDGs. This report examines the level of achievement of these goals in Morocco in 2020 and outlines the prospects for their evolution by 2030 (HCP 2020). This work highlights a participatory approach between government departments, national institutions, public entities, socio-professional organizations, and non-governmental organizations within the National Commission for Sustainable Development to illustrate their collaborative efforts. The report emphasizes the importance of a stable institutional structure to ensure impartiality and inclusivity, highlighting its objective to strengthen communication efforts aimed at enhancing ownership of the SDGs in all areas of national and territorial governance. The report represents the state of all key sustainability goals, among which is "end hunger, achieve food security, improve nutrition, and promote sustainable agriculture," serving as an authoritative resource and presenting the latest information and perspectives regarding the agricultural sector in Morocco. The report contributes to a broader understanding of Morocco's journey towards achieving the aforementioned SDGs by carefully examining its current agricultural status.

The report represents several progressive and monitoring policies undertaken to revitalize and implement the main strategies and programs aimed at achieving the optimal and desirable level of sustainability of Moroccan agriculture. It highlights the proactive approach of the Moroccan government in implementing a new policy aimed at reforming land policies, specifically to give women in rural areas the opportunity to use collective agricultural lands designated as "Soulaliyate." This strategic initiative aims to promote small-scale farmers and foster sustainable agricultural practices. Notably, in 2019, three laws were enacted to protect the rights of Soulaliyate women regarding the use and ownership of communal lands. These laws, namely Law No. 62.17, Law No. 63.17, and Law No. 64.17, respectively address the administrative guardianship of Soulaliyate communities, the demarcation of Soulaliyate lands, and the regulation of Soulaliyate lands located in irrigated areas. These legislative measures aim to ensure the active participation of Soulaliyate women in agricultural activities and to secure their legitimate access to these lands (Real 2021).

In addition to this crucial strategy, several central practical strategies aim to enhance the value of production, improve processing units, and strengthen the marketing tactics of agricultural products. These strategies are considered crucial for stimulating the growth of this vital sector in the country. Achieving these three key strategic points involves diverse methods that vary from one country to another. However, accomplishing these objectives requires a multifaceted and synergistic approach involving a confluence of policy interventions, technological advancements, and institutional mechanisms. Before delving into a detailed explanation of recent and ongoing projects aimed at achieving these key strategic points to promote the agricultural sector in Morocco, it is essential to briefly highlight how countries generally manage to achieve these objectives. Specifically, what are the instruments or means actually used to make them more practically achievable?

## 7 Navigating the Pathway to Progress: A Thorough Analysis of the Outcomes Achieved Through Moroccan Government Programs

This section presents the outcomes of Moroccan efforts and projects aimed at developing agricultural industrialization for peace and sustainability. It also discusses the results of these government programs, demonstrating the significant progress and economic benefits achieved. The analysis examines how Morocco is leveraging internal drivers, such as technology adoption, infrastructure development, human capital improvement, and institutional support, to foster endogenous growth and resilience in the agricultural sector.

The first program that Morocco recently implemented for agricultural industrialization for peace, sustainability, and sector development is PADIDZAR. This initiative, approved on October 27, 2021, and signed on November 23, 2021, aims to improve agricultural practices and promote rural development in specific geographical areas. The program falls under the sovereign sector of Agriculture and Rural Development, with a DAC sector code of 31,110. The scheduled completion date for PADIDZAR is December 31, 2026, and it is classified under Environmental Category 2. The commitment for the project is 93,597,636.05 UA, and it is currently in the implementation phase, actively working towards achieving its objectives. It is funded by the AfDB, which has granted a loan of $95 million (equivalent to 114.03 million euros) to the Kingdom of Morocco to finance the Support Program for Inclusive and Sustainable Agricultural and Rural Development (PADIDZAR) (AfDB 2021). The AfDB loan aims to align this project with its commitment to promoting inclusive and sustainable development in Morocco's agricultural and rural areas, thus supporting the country's goals of reducing poverty and promoting inclusion in these regions. The program is structured as a results-based financing enterprise in line with the implementation of the Green Generation Strategy (SGV) 2020–2030, validated in January 2020, to strengthen the National Program for Drinking Water Supply and Irrigation 2020–2027, also approved in January 2020. According to findings presented in a report by the High Commission for Planning (HCP), the incidence of poverty in rural geographical areas is twice the national average, with 79.4% of individuals classified as being in poverty residing in this specific region (El Bazzim 2022). The project is committed to achieving several key objectives by 2030. These objectives include doubling agricultural GDP and agricultural exports, achieving a twofold increase in water efficiency, establishing 400,000 middle-class households while extending social protection to 3–4 million farmers, developing one million hectares of collective farms involving 350,000 new farmers and entrepreneurs, and training 150,000 youth. Additionally, the project aims to increase the value of 70% of agricultural production by 2030.

In general, the program will contribute to the achievement of Objective 4 of the National Sustainable Development Strategy, which aims to accelerate the implementation of the national policy to combat climate change. It helps to achieve the objectives of the SNDD 2030 by working with regions to combat climate change and improve climate governance, in line with their strategic goals (AfDB 2021). Therefore, managing upstream water intake and reducing losses during its distribution in irrigation channels can help ensure water availability in agricultural plots and improve agricultural productivity and production in small and medium irrigation areas. This will help mitigate the effects of climate change, such as increasingly frequent and prolonged variations in precipitation and drought episodes.

To ensure the success of the projects and programs, the government of Morocco, over the past two years, has undertaken a series of analytical studies that have played a crucial role in shaping the current operation. These studies are described as follows. The first study, titled "Opportunities for Rural Agro-Industrial Development in Morocco: The Example of Rural Agro-Industrial Zones (RAZs)," aimed to identify opportunities for creating platforms that promote the development of agricultural products in rural areas, thereby fostering rural agro-industrial growth in Morocco. The initial phase of this study focused on analyzing opportunities within the agricultural sub-sectors of the pilot production basins Loukkos and Doukkala, located in the regions of Tanger-Tétouan-Al Hoceima and Casablanca-Settat, respectively.

The second study, titled "Management of the COVID-19 Pandemic: Impact Study on the Moroccan Agro-Industry and Proposal of a Revival Strategy," was conducted to assess the socio-economic impact of the COVID-19 pandemic on the agri-food industry in Morocco. It specifically examined the industry's reliance on imports and exports. Based on a comprehensive impact analysis, this study formulated policy recommendations aimed at mitigating the effects of the COVID-19 pandemic on the agri-food sector in the medium and long term.

These analytical studies have significantly contributed to the informed design of the current operation by providing valuable insights into opportunities for rural agro-industrial development and proposing strategies to address the challenges posed by the COVID-19 pandemic within the Moroccan agri-food industry. Their conclusions and recommendations have played a crucial role in defining the operational approach and objectives.

In addition to the aforementioned projects supported by the AfDB to promote agricultural development in Morocco, there is another project called the Competitive and Resilient Cereal Crop Development Support Program. This rural and agricultural development project was approved by the Moroccan government with total funding of $162,227,820.03. The anticipated completion date is December 31, 2026, with the final payment scheduled for December 31, 2024. The $162 million project funding demonstrates the Moroccan government's commitment to advancing the agricultural sector and improving rural well-being. If successfully implemented, the project has the potential to significantly enhance agricultural productivity, rural incomes, and livelihoods, as its primary objective is to improve agriculture and the quality of life for rural residents. The project addresses constraints faced by rural communities and farmers in Morocco through interventions targeting infrastructure, technology, skills, and market access. With proper implementation, the project can make a significant contribution to agricultural productivity and rural well-being. The Moroccan government has formally requested a financial injection of €199 million from the African Emergency Food Production Fund (AEPF) to support the implementation of the Competitive and Resilient Cereal Crop Development Support Program (CRCCDSP) between 2022 and 2024 (AfDB 2023). The overall objectives of the CRCCDSP initiative encompass two main aspects: firstly, addressing pressing challenges of drought and food price hikes by accelerating cereal production and its immediate availability; and secondly, enhancing the resilience and competitive capacity of the cereal sector in the Moroccan agricultural landscape. The main objectives of the CRCCDSP program are as follows:

- Increase cereal production by at least 10% through support in inputs, investments in irrigation infrastructure, agricultural research, and extension services.
- Reduce Morocco's dependence on cereal imports by improving the productivity and resilience of domestic cereal production.
- Improve the policy and institutional framework of the cereal sector to strengthen its long-term competitiveness and resilience to climate change.

The program is expected to benefit approximately 980,000 cereal farmers, including 70,000 women. This will also improve food security and accessibility for the Moroccan population. Additionally, private sector actors involved in the cereal value chain will reap the benefits of this support (AfDB 2023). In summary, the CRCCDSP program aims to address immediate food security needs in Morocco through emergency measures while establishing a sustainable and resilient cereal sector through investments, research, and institutional reforms. With proper execution, this program could significantly contribute to food security and rural livelihoods in Morocco. However, challenges related to smallholders, climate impacts, and policy weaknesses need to be addressed and mitigated to ensure the program's success.

## 8 Discussion

This study leverages the theory of endogenous growth to analyze the internal dynamics of Morocco's agricultural sector, highlighting the role of structural transformation in enhancing resilience against external shocks. Endogenous growth theory emphasizes the importance of internal drivers, such as technological innovation, human capital development, and institutional support, in fostering economic growth within a specific industry or sector. This theoretical framework is particularly relevant to Moroccan agriculture, a sector employing over 40% of the country's workforce and playing a crucial role in the national economy, food security, and socio-economic stability, all of which are foundational to promoting peace and sustainability.

The application of endogenous growth principles can help Morocco's agricultural sector withstand external pressures, including the COVID-19 pandemic, the Russia-Ukraine conflict, and climate change-induced droughts. The study identifies key internal drivers that can be harnessed to promote growth, resilience, peace, and sustainability. Promoting research, development, and the adoption of advanced technologies, such as drought-resistant seeds, efficient irrigation systems, and digital agricultural tools, can significantly boost productivity and farmer incomes (Moussaid et al. 2023a, b). Investment in agricultural R&D by the government and international donors is vital to this effort. Enhanced productivity and incomes contribute to economic stability, reducing the likelihood of conflict and promoting social cohesion.

Expanding rural infrastructure, including roads, storage and processing facilities, and access to affordable credit and inputs, can better connect farmers to markets and value chains, creating opportunities for growth. Improved infrastructure fosters economic development and reduces disparities, which are crucial for maintaining peace and ensuring sustainable development. Enhancing farmers' access to extension services, technical education, and skills development can enable them to sustainably intensify and diversify their production, contributing to overall sectoral growth. Educated and skilled farmers are better equipped to manage resources sustainably, thus supporting long-term environmental stewardship and economic stability, essential for peaceful societies.

Strengthening farmer associations, agricultural policy frameworks, and financial services tailored to smallholders can empower local agricultural systems and foster endogenous growth. Robust institutions provide a framework for equitable resource distribution and conflict resolution, essential for sustaining peace and promoting equitable development. By exploiting these internal and structural drivers, Moroccan agriculture can unlock its full growth potential despite mounting external pressures. The theory of endogenous growth provides a pertinent framework for promoting resilience, peace, and sustainability within the sector.

The Green Morocco Plan, a significant governmental initiative, aims to enhance the agricultural sector by improving legislative, institutional, and regulatory frameworks to attract international investors and improve market access for farmers. This plan also emphasizes sustainable environmental management, aiming to make the sector more competitive and productive. These initiatives have led to substantial improvements in market access, financial resources, and overall competitiveness, thereby contributing to economic stability and social peace. The Green Generation Strategy (GGV) 2020–2030, introduced by King Mohammed VI, focuses on the economic integration of rural youth, enhancing marketing efficiency, and promoting environmental sustainability. This strategy addresses immediate and long-term economic challenges exacerbated by the COVID-19 pandemic, such as drought and economic disruptions. The GGV aims to expand the rural middle class, create jobs, and develop skills, driving significant transformation in rural economies. By improving economic opportunities and living standards in rural areas, the GGV contributes to reducing poverty and preventing conflict, thereby promoting peace and sustainability.

Programs like PADIDZAR and the Competitive and Resilient Cereal Crop Development Support Program demonstrate Morocco's commitment to inclusive and sustainable agricultural development. PADIDZAR, supported by the AfDB, focuses on improving agricultural practices, enhancing water management, and fostering rural development. It aims to double agricultural GDP and exports by 2030, improve water efficiency, and create substantial middle-class households among rural farmers. Economic empowerment of rural communities through such initiatives promotes stability and peace. The Competitive and Resilient Cereal Crop Development Support Program aims to increase cereal production by at least 10% and reduce Morocco's dependence on cereal imports, benefiting nearly one million cereal farmers, including a significant number of women. This program addresses the challenges of drought and food price hikes by boosting cereal production and enhancing sector resilience and competitiveness. By ensuring food security and economic resilience, the program supports sustainable development and peace.

The Moroccan government has enacted crucial legislative measures to support these initiatives, such as laws protecting the rights of women (Soulaliyate) to use and own communal lands. These reforms ensure active participation in agricultural activities and secure land rights for rural women, promoting gender equality in the agricultural sector. Empowering women and ensuring equitable resource distribution are key components of peacebuilding and sustainable development. The World Bank's adaptive approach in Morocco has ensured alignment with the changing needs and priorities of the agricultural sector. Programs like the Irrigated Agriculture Modernization Project have evolved to incorporate climate adaptation techniques, reflecting the flexibility and responsiveness of international support to Morocco's agricultural development. International cooperation and support play a vital role in sustaining peace and promoting long-term sustainability.

Despite the substantial progress achieved, challenges remain, including addressing the impacts of climate change, enhancing smallholder productivity, and ensuring the sustainability of agricultural practices. Future strategies must focus on these areas to build on current successes and ensure long-term resilience and growth. Addressing these challenges is critical for maintaining social stability and promoting sustainable development. Overall, the Moroccan government's efforts, supported by international institutions like the AfDB and the World Bank, have significantly advanced the agricultural sector. These initiatives have not only promoted economic growth but also contributed to social and environmental sustainability. The comprehensive strategies and their evolution over time provide valuable insights into the potential for agricultural industrialization to foster peace and sustainability in Morocco. Ongoing and future projects will continue to build on these achievements, aiming for a prosperous, peaceful, and sustainable agricultural sector.

To further enhance Morocco's agricultural sector and align with the Sustainable Development Goals (SDGs), particularly those related to agriculture and environmental sustainability, It is recommended to utilize Awqaf-led green finance as a strategic tool to further enhance Morocco's agricultural sector and align with the Sustainable Development Goals (SDGs), particularly those related to agriculture and environmental sustainability. Awqaf, a form of Islamic endowment used for charitable purposes, can be effectively harnessed for green finance projects, especially those focused on agriculture. Awqaf-led green finance can support a variety of agricultural initiatives, such as investment in sustainable agricultural practices, development of rural infrastructure, support for research and development, and empowering farmers through education and training (Zain et al. 2024). By directing Awqaf resources towards these areas, Morocco can not only advance its agricultural sector but also contribute to broader goals of environmental sustainability and socio-economic stability. This approach aligns with the SDG targets, fostering a more resilient, productive, and equitable agricultural system that supports peace and sustainable development.

## 9 Conclusion

In conclusion, the Moroccan government has made significant strides in promoting the agricultural sector, particularly through the Green Morocco Plan and the Green Generation Strategies. However, additional investments in innovation, industrialization, and effective marketing tactics are needed to further increase productivity and profitability. The ongoing projects and initiatives, in collaboration with international institutions, hold significant potential. A more thorough evaluation of their impact is necessary to ensure they effectively contribute to the sustainable growth of the Moroccan agricultural sector. Adopting a balanced approach that combines product valorization strategies, marketing tactics, and innovative projects could enable the Moroccan agricultural sector to achieve its full potential and contribute optimally to the national economy.

## References

Abdelmajid S, Mukhtar A, Baig MB, Reed MR (2021) Climate change, agricultural policy and food security in Morocco. In: Behnassi M, Barjees Baig M, El Haiba M, Reed MR (eds) Emerging challenges to food production and security in Asia, Middle East, and Africa. Springer International Publishing, pp 171–196. https://doi.org/10.1007/978-3-030-72987-5_7

ADBG (n.d.) Morocco—MapAfrica, agriculture and rural development projects. Retrieved 11 Aug 2024, from https://mapafrica.afdb.org/en/countries/MA

Adghough J (2018) Economic overview of agriculture and food in Morocco. https://cuvillier.de/en/shop/publications/7879-economic-overview-of-agriculture-and-food-in-morocco

AfDB (2021) Morocco—inclusive and sustainable development project for agricultural and rural areas (PADIDZAR) [Result-based financing operation appraisal report]. https://www.afdb.org/sites/all/libraries/pdf.js/web/viewer.html?file=https%3A%2F%2Fwww.afdb.org%2Fsites%2Fdefault%2Ffiles%2Fdocuments%2Fprojects-and-operations%2Fmorocco_-_inclusive_and_sustainable_development_support_programme_for_agricultural_and_rural_areaspadidzar_-_appraisal_repport.pdf#page=1&zoom=auto,-13,848

AfDB (2023) African development bank group annual report [text]. African Development Bank Group. https://www.afdb.org/en/documents/annual-report-2022

Ait Ali A, Dadush U, Mengoub FE, Tsakok I (2022) The Russia-Ukraine war and food security in Morocco [policy brief]. https://www.policycenter.ma/publications/russia-ukraine-war-and-food-security-morocco

Arndt C, Tyner WE (2003) Policy and progress in Moroccan agriculture: a retrospective and perspective. In: Food, agriculture, and economic policy in the Middle East and North Africa. Emerald Group Publishing Limited, pp 231–256. https://doi.org/10.1016/S1094-5334(03)05014-3/full/html

Asedrem S (2021) The green Morocco plan: a case study of the unintended consequences of sustainable development initiatives. https://stars.library.ucf.edu/etd2020/470/

Ayham T (2022) Morocco: drought assessment report brief. CARE International. https://www.careevaluations.org/wp-content/uploads/Drought-Assessment-Report-CARE-Morocco.pdf

Barry NY, Dia K, Ndoye A, Ly R (2023) Ukraine crisis brief series wheat production outlook in Morocco amid the Ukraine crisis. https://www.academia.edu/download/102252180/ucbs.pdf

Bertin S, Brouillard J, Doffonsou RA, Weiss C, Martinez I (2021) Private sector developement in Morocco challanges opportunities in time of covid-19 [joint report]. https://www.eib.org/attachments/country/private-sector-morocco-covid.pdf

Bouramdane A-A (2024) Morocco's path to a climate-resilient energy transition: identifying emission drivers, proposing solutions, and addressing barriers. Sci Technol Energy Transition 79:26

Chafil R (2018) National strategy for sustainable development 2030. The Global Water Operators' Partnerships Alliance (GWOPA). http://www.gwopa.org/sites/default/files/2023-06/4_National_Strategy_for_Sustainable_Development_and_Territorialization_of_its_Implementation_in_Morocco.pdf

Diaz Cassou J, Iraqi A, Megevand C, Marzo F (2022) Morocco economic update: the recovery is running dry. World Bank Group, Washington, D.C. https://documents.worldbank.org/en/publication/documents-reports/documentdetail/099529307192239926/IDU020d792070df4a04e830957f0f9fd56c0f71e

El Bazzim R (2022) COVID 19 and poverty in Morocco. https://doi.org/10.1007/978-3-030-68127-2_319-1

El Houssain B, Fadlaoui A, Allali K, Arrach R (2019) Contract farming in the Morocco cereal sector: contract clauses, ambiguity, and opportunism. Development 1:2

Erraji-Morocco World News, A. (2024) Morocco imported 25% of wheat from Russia in December 2023. https://www.moroccoworldnews.com/2024/01/359930/morocco-imported-25-of-wheat-from-russia-in-december-2023

Ghanem H (2015) Agriculture and rural development for inclusive growth and food security in Morocco. Brookings global working paper series. https://www.brookings.edu/wp-content/uploads/2016/07/Agriculture_WEB_Revised.pdf

HCP (2020) Examen National Volontaire de la Mise en Oeuvre des Objectifs de Développement Durable [Examen National Volontaire de la mise en œuvre des ODD, Rapport National]. https://www.wmaker.net/testhcp/file/229720/

HCP (2023) La situation du marché du travail en 2022. Enquête nationale sur l'emploi, Haut-Commissariat au Plan. https://www.hcp.ma/La-situation-du-marche-du-travail-en-2022_a3661.html

IBRD & WBG (2020) International bank for reconstruction and development program appraisal document on a proposed loan in the amount of EUR214.2 Million (US$250 million equivalent) to the Kingdom of Morocco for a Morocco green generation program-for-results [IBRD-IDA world bank group report]. https://documents1.worldbank.org/curated/en/245801608346893390/pdf/Morocco-Green-Generation-Program-for-Results-Project.pdf

ITA (2022) Morocco—country commercial guide—water. International Trade Administration. https://www.trade.gov/country-commercial-guides/morocco-water

Kydd J, Thoyer S (1992) Structural adjustment and Moroccan agriculture: an assessment of the reforms in the sugar and cereal sectors. https://www.oecd-ilibrary.org/content/paper/217714011805

Mattia T (2021) Promoting investment climate reforms in Morocco's agri-food sector. Policy Commons. https://policycommons.net/artifacts/3838187/promoting-investment-climate-reforms-in-moroccos-agri-food-sector/4644079/

Michel-Kerjan E, Scawthorn C, Baeumler AE, Banerjee A, Rondot P, Medouar M, Mahul O, Boudreau L, Davila-Bonazzi A, Dana J (2014) Building Morocco's resilience: inputs for an integrated risk management strategy. The World Bank

Montanari B (2012) The future of agriculture in the high atlas mountains of Morocco: the need to integrate traditional ecological knowledge. https://doi.org/10.1007/978-3-642-33584-6_5

Moussaid F, Bachegour H, Jerry M, Qafas A (2023a) Enhancing resilience in food systems: a comprehensive review of innovative measures. https://doi.org/10.5281/zenodo.8299459

Moussaid F, Jerry M, Qafas A (2023b) Examining the nexus water demand-economic growth in Morocco: a tapio decoupling index analysis. Int J Account Fin Audit Manage Econ. https://doi.org/10.5281/zenodo.8388151

Office of Agricultural Affairs, U. S. E. R. (2024) Morocco—country commercial guide: agricultural sector. U.S. Department of Commerce. https://www.trade.gov/country-commercial-guides/morocco-agricultural-sector

Oxford Business Group (2020) The report Morocco 2020. https://www.morocconow.com/wp-content/uploads/2021/10/OBG-The-Report-Morocco-2020.pdf

Rahhou J (2023) Morocco's agricultural production remains limited in 2023. Morocco World News. https://www.moroccoworldnews.com/2023/04/354925/moroccos-agricultural-production-remains-limited-in-2023

Real C (2021) L'influence du mouvement revendicatif des femmes Soualaliyates sur l'adoption de politiques foncières égalitaires au Maroc. Institut du Genre en Géopolitique. https://igg-geo.org/2021/04/26/linfluence-du-mouvement-revendicatif-des-femmes-soulaliyates-sur-ladoption-de-politiques-foncieres-egalitaires-au-maroc-2-2/

Statista Research Department (2024a) Morocco: value added by agriculture 2023. Statista. https://www.statista.com/statistics/1230909/value-added-by-agriculture-in-morocco/

Statista Research Department (2024b). Agriculture in Morocco—statistics & facts. Statista. https://www.statista.com/topics/7565/agriculture-in-morocco/

The Agency for Agricultural Development (ADA) (2020) His majesty King Mohammed VI launches the new agricultural strategy "generation green 2020–2030." ADA. https://www.ada.gov.ma/en/news/his-majesty-king-mohammed-vi-launches-new-agricultural-strategy-generation-green-2020-2030

The World Bank (2022) Morocco economic update the recovery is running dry. International Bank for Reconstruction and Development, The World Bank. https://documents1.worldbank.org/curated/en/099529307192239926/pdf/IDU020d792070df4a04e830957f0f9fd56c0f71e.pdf

UNDP (2017) National adaptation plans in focus: lessons from Morocco. United Nations Development Programme. https://www.undp.org/sites/g/files/zskgke326/files/publications/Morocco_NAP_country_briefing.pdf

UNFCCC (2023a) Morocco's experience of loss and damage management and finance: the case of drought, Transitional Committee workshop on loss and damage. The United Nations Framework Convention on Climate Change (UNFCCC). https://unfccc.int/sites/default/files/resource/Loss%20and%20damage%20MTEDD.pdf

UNFCCC (2023b) Report of the global environment facility to the twenty-eighth session of the conference of the parties to the united nations framework convention on climate change.

United Nations Framework Convention on Climate Change. https://unfccc.int/sites/default/files/resource/cp2023_06E.pdf

WBG (n.d.) The fund to address loss and damage, World Bank group [text/HTML]. World Bank. Retrieved 8 Aug 2024, from https://www.worldbank.org/en/programs/funding-for-loss-and-damage

World Bank (2023) The World Bank Group's engagement in Morocco, fiscal years 2011–21 country program evaluation. Independent Evaluation Group, World Bank, Washington, DC. https://www.bing.com/search?EID=MBSC&form=BGGCMF&pc=U707&DPC=BG02&q=World+Bank+(2023).+The+World+Bank+Group's+engagement+in+Morocco%2C+Fiscal+Years+2011–21.+Country+Program+Evaluation.+Independent+Evaluation+Group.+Washington+DC%3A+World+Bank

World Bank Group (2022) Morocco's economic update—April 2022. World Bank. https://www.worldbank.org/en/country/morocco/publication/economic-update-april-2022

Zain MNR, Tounchibine O, Noor AM, Lechheb H (2024) Awqaf-led green finance a technical know-how. Islamic Green Finance. https://www.taylorfrancis.com/chapters/edit/10.4324/9781032672946-21/awqaf-led-green-finance-razinah-mohd-zain-oumaima-tounchibine-azman-mohd-noor-houda-lechheb

# The Evolution of Environmental Diplomacy: Reviewing Four Decades of Research

Faribo Idibekzoda

**Abstract** The development of international environmental cooperation, the introduction of new tools for environmental protection, and the popularization of "ecological" behavior have all contributed to the emergence of environmental diplomacy. Despite its relevance in international relations, environmental diplomacy is characterized by diverse and fragmented literature. Amidst pressing global environmental issues, it is important to understand how environmental diplomacy is interpreted and adopted. Employing a systematic review of 744 scholarly articles, this paper synthesizes the main research themes and trends in the field of environmental diplomacy. The findings reveal significant shifts in the perception of environmental diplomacy, highlighting its effectiveness in multilateral agreements and its intricate links with peace and equality. Moreover, the study identifies major challenges, such as power imbalances and implementation issues, but also underscores the positive impact of environmental diplomacy on raising environmental awareness and fostering international cooperation. This research explores the evolution and impact of environmental diplomacy, enhancing our understanding of how international relations intersect with environmental management and offering insights for strengthening global environmental governance.

**Keywords** Environmental diplomacy · Environmental cooperation · Peacebuilding · Sustainability · Sustainable development goals (SDGs)

F. Idibekzoda (✉)
International Peace and Coexistence Program, Graduate School of Humanities and Social Sciences, Hiroshima University, Higashihiroshima, Japan
e-mail: m226987@hiroshima-u.ac.jp

# 1 Introduction

Pressing global environmental issues, such as climate change, pollution, ocean acid-ification, and deforestation, among others, have prompted states and international organizations to develop diplomatic tools for environmental cooperation. Environ-mental diplomacy emerged as a key concept to address shared environmental chal-lenges. It involves conducting international relations by promoting a shared commit-ment to the conservation of natural resources through sustainable operation and responsible environmental management (Mcintire 2014).

Addressing environmental challenges presents numerous complexities due to their transboundary nature and the varying capacities of nations to respond. These challenges include the imperative to harmonize economic development with envi-ronmental preservation, bridge the gap between developed and developing nations, and manage the diverse interests of multiple stakeholders. For example, addressing climate change necessitates global cooperation, but disparities in economic priori-ties and levels of industrialization make consensus-building a complex task. Inter-national environmental diplomacy has been instrumental in addressing these chal-lenges through collaborative efforts and multilateral agreements. Notable examples include the Paris Agreement, which aims to mitigate global warming by setting emission reduction targets for countries, and the Convention on Biological Diver-sity, which focuses on conserving biodiversity and promoting sustainable natural resource utilization. Another key example is the Montreal Protocol, which effectively addressed ozone layer depletion by phasing out the production of ozone-depleting substances. Additionally, the Ramsar Convention emphasizes the preservation and sustainable use of wetlands, highlighting the critical importance of safeguarding essential ecosystems. These endeavors underscore the significance of diplomatic negotiations and cooperative frameworks in overcoming the obstacles to effective environmental governance.

This chapter provides a systematic approach to understanding this broad and complex concept of environmental diplomacy. Environmental diplomacy encom-passes a wide array of related terms and practices, including environmental cooper-ation, environmental negotiations, climate cooperation, climate diplomacy, ecolog-ical cooperation, and ecological diplomacy. This diversity illustrates the broadness of environmental diplomacy and the challenge of synthesizing a cohesive body of knowledge. While there is a significant body of work on environmental diplomacy, the literature is often fragmented across disciplines, such as political science, envi-ronmental science, and international relations, lacking a cohesive framework to inte-grate these perspectives. This fragmentation results in knowledge gaps and incon-sistent methodologies that limit comprehensive analysis and policy development. Therefore, this paper contributes to the literature by offering a systematic and inte-grative review of environmental diplomacy. It aims to synthesize existing research, identify common themes and gaps, and propose a structured approach to studying environmental diplomacy.

In the following section, I present the bibliometric analysis of selected articles as an overview of the literature landscape of environmental diplomacy. And then, the section on the evolution of environmental diplomacy is divided into four decades to outline the developments in the study of environmental diplomacy. The bibliometric analysis sets the stage by providing a broad overview of the research landscape, while the thematic analysis offers detailed insights into the content and direction of the research. Together, they allow for a holistic assessment of the current state of the scientific landscape of environmental diplomacy, identifying areas for future research, and informing policies and practices in environmental diplomacy. This chapter concludes with key findings and recommendations for future research.

## 2 The Literature Landscape of Environmental Diplomacy

Scholars provide varying definitions of environmental diplomacy based on their research objectives and questions. In general, environmental diplomacy refers to the practice of negotiating and managing international agreements and policies aimed at addressing global environmental issues. It involves formal negotiations, treaty-making, and diplomatic interactions to establish legal frameworks and resolve environmental conflicts (Chasek 2012). It emphasizes the role of states, international organizations, and non-state actors in crafting and implementing environmental treaties and protocols (Betsill and Corell 2008). The role of non-state actors, such as NGOs and the private sector, in environmental diplomacy is increasingly recognized. These actors contribute to shaping negotiations and outcomes, although their influence varies depending on the context and issue (Betsill and Corell 2008). As defined by scholars like Wapner (1995), environmental diplomacy involves the strategic engagement of states, international organizations, and non-state actors in negotiating and managing agreements to address global environmental challenges. Wapner highlights the importance of both traditional state-centric diplomacy and the growing influence of non-state actors in shaping international environmental policies (Wapner 1995). Wapner also broadens the scope of environmental diplomacy by emphasizing the complex interplay between various actors and the multi-layered nature of environmental governance. This comprehensive view is crucial for understanding how diverse interests and power dynamics shape the outcomes of environmental negotiations and policies.

Environmental diplomacy is closely related to environmental cooperation but differs in focus and application. Collaborative environmental efforts strongly emphasize working together among nations, organizations, and communities to address mutual environmental challenges. This approach typically involves joint initiatives, sharing of information, and building capacity to achieve sustainable development goals. For instance, global initiatives such as the United Nations' Sustainable Development Goals (SDGs) promote international partnerships to address pressing issues like climate change and biodiversity loss (UN 2015). Conversely, environmental diplomacy is the practice of negotiating and managing international environmental agreements and policies, including formal negotiations, treaty-making, and diplomatic interactions to resolve environmental conflicts and establish legal frameworks. Environmental diplomacy is exemplified by major international agreements such as the Paris Agreement, which aims to limit global warming through legally binding commitments (UNFCCC 2015). The primary distinction lies in their operational mechanisms: cooperation relies on collective action and shared efforts, while diplomacy focuses on formal negotiation and policy-making. Both play critical roles in global environmental governance but operate at different levels and through distinct processes (Biermann 2007). Effective environmental governance requires integrating cooperation and diplomacy to address the multifaceted nature of global environmental challenges.

For the selection of the bibliographic portfolio, I used the *Web of Science* for its comprehensive database. Using the research topic "environmental diplomacy," I applied the inclusion criteria for research articles in English and Russian.[1] Environmental diplomacy, however, is related to other terms, so the search string also included words such as environmental cooperation, environmental negotiations, climate cooperation, climate diplomacy, ecological cooperation, ecological diplomacy, green diplomacy, green cooperation, water diplomacy, and water cooperation. Only document types such as articles were used to avoid duplication of publications and repetitions of findings. This step resulted in 926 articles as of September 29, 2023.

To narrow down the bibliographic portfolio, *Web of Science* categories relevant to environment and diplomacy were selected, such as international relations, political science, history, economics, environmental studies, environmental science area studies, geography, law, history, and water resources. This step resulted in 744 articles used for bibliometric analysis. No specific timeframe was selected during the selection process in order to trace the earlier conceptualization of environmental diplomacy. Figure 1 shows an upward trend in publications related to environmental diplomacy since 1972.

The study of environmental diplomacy is interdisciplinary. Among the selected disciplines based on the *Web of Science* categories, Environmental Studies has the most publications, with 21% of the articles, followed by Political Science and International Relations at 16% and 15%, respectively (Fig. 2).

The geographic distribution (Fig. 3) of authors' institutional affiliation indicates Western dominance (except for China with 83) in the discourse of environmental diplomacy. The United States leads with 201 articles, followed by the UK (n = 83), Germany (n = 70), the Netherlands (n = 55), Canada (n = 41), Sweden (n = 38), Australia (n = 36), Norway (n = 32), and Switzerland (n = 30). The disparity in research publications between the Global North and the Global South is not surprising, reflecting the dominance of Global North countries in academic publishing. Additionally, the rise of environmental diplomacy is closely linked to

---

[1] The author is fluent in English and Russian languages.

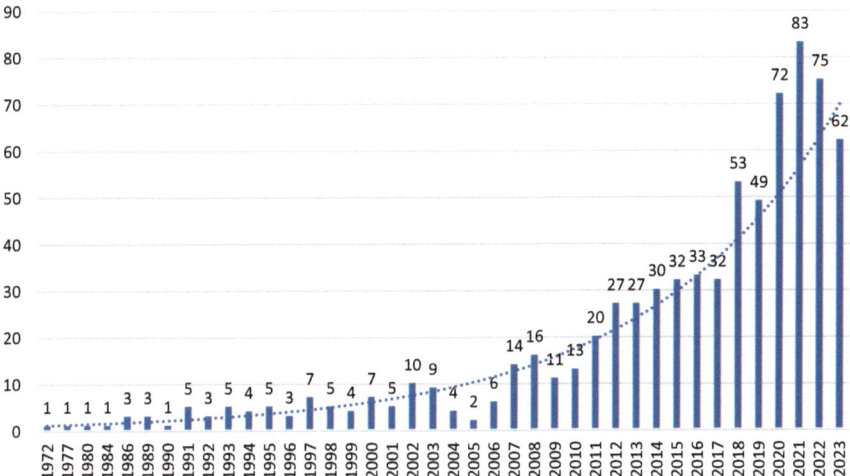

**Fig. 1** Number of published articles annually

**Fig. 2** Top 10 web of science categories with the most publications on environmental diplomacy

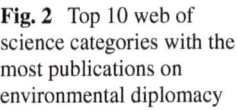

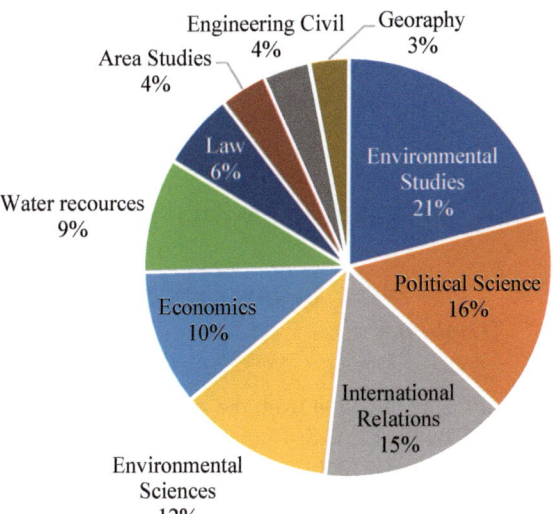

issues arising from industrialization. The countries in the Global North, which industrialized earlier, have had more time and resources to invest in research and publications on environmental issues. On the other hand, many countries in the Global South are still dealing with industrialization and its associated challenges, limiting their capacity to produce a comparable volume of research on environmental diplomacy.

Conducting a bibliometric analysis serves as a helpful tool for researchers to map out the intellectual structure of the field. It provides a quantitative overview of the research landscape and identifies the prominent disciplines and the geographical

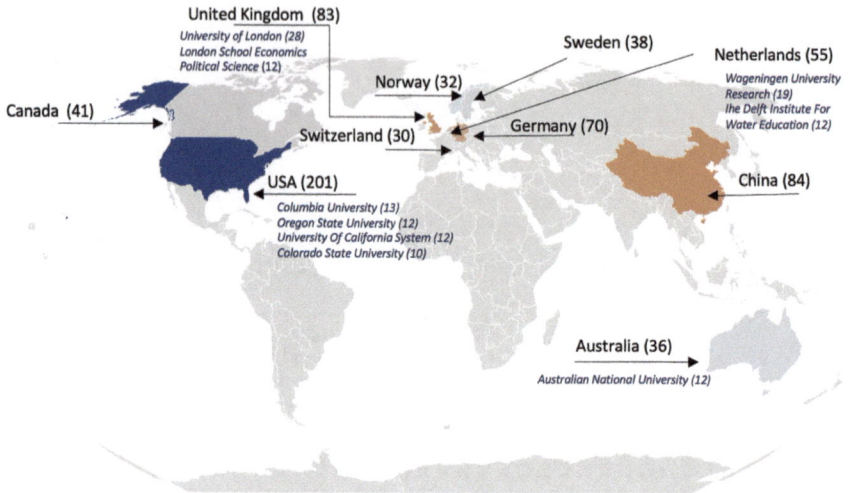

**Fig. 3** Top 10 countries and top 10 institutions with the most published research articles related to environmental diplomacy

distribution of research outputs. The following thematic analysis delves deeper into the research content to identify recurring themes, concepts, and trends.

## 3 The Evolution of Environmental Diplomacy

### 3.1 1980–1990: Narrow Conceptualization of Environmental Diplomacy

The earliest publication from the selected bibliographic portfolio is Bilder's (1972) paper on the 1971 U.S.-Canada Great Lakes Water Quality Agreement and its historical context, highlighting its insights for international environmental cooperation and underscoring the importance of bilateral and regional collaboration in managing complex environmental problems while acknowledging challenges and potential benefits. The paper discusses the U.S.-Canadian experience in addressing Great Lakes pollution, emphasizing cooperation needs, legal frameworks, institutions, objectives, burden-sharing, and coordination (Bilder 1972). Throughout the paper, including the discussion on the efficacy of localized approaches, the role of informal cooperation, and the value of expert institutions, Bilder used the term "environmental cooperation" rather than "environmental diplomacy."

The term "environmental diplomacy" first appeared in 1983 in the book by John Edward Carroll "Environmental Diplomacy: An Examination and a Prospective of Canadian—U.S. Transboundary Environmental Relations," which provides insight into US-Canadian environmental relations from the late 1800s (Thompson 1984). From today's perspective, this book might seem outdated, and issues are now resolved, but particularly after publication, the term became widely used and paved the way for future discussions.

Environmental diplomacy is also described in Mumme's 1984 article about the US-Mexico transboundary environmental issues. While it does not explicitly define environmental diplomacy, it provides insights into what it entails, referring to the diplomatic strategies and tactics one country uses to address and resolve environmental disputes with another country. This involves negotiations, discussions, and interactions between the two countries aimed at finding solutions to environmental challenges that affect both nations, particularly those along the shared border. Mumme highlights the importance of understanding whether this relationship is characterized by dependency or interdependence, as it has implications for how environmental disputes between the two countries are resolved (Mumme 1984). In terms of the US-Mexico relationship, if there is a dependency, the United States might use its economic strength to influence outcomes, potentially including the use of threats or sanctions. However, if interdependence is more prevalent, diplomatic strategies should be focused on finding cooperative solutions, disentangling issues, or providing positive incentives for Mexican participation in transboundary environmental agreements (Mumme 1984).

Most of the earlier articles on environmental diplomacy are devoted to US environmental cooperation with its neighbors (Bilder 1972; Sinclair 1986; Carroll 1986). The complex issue of transboundary water management between the US and Canada and the constant presence of the issue on the diplomatic agenda of both states is described by Bilder (1972) and Carrol (1986), while Sinclair (1986) analyzed the efforts to address the environmental challenges in the border region through the 1983 US-Canada agreement. While it outlined a general framework for cooperation, it did not provide detailed plans or mechanisms for how the two countries would tackle specific environmental challenges. This lack of specificity could have hindered the practical effectiveness of the agreement.

The conceptualization of environmental diplomacy during this period was informed by the recognition of interdependence and shared responsibility for environmental protection. The abovementioned studies emphasized that effective environmental diplomacy required more than formal agreements; it necessitated ongoing dialogue, mutual understanding, and adaptive management strategies. However, the narrow focus of early environmental diplomacy scholarship, primarily on Canada and the United States, limited the geographic scope of analyses and concentrated on bilateral or regional issues rather than global diplomatic efforts. Despite this, these studies laid the groundwork for broader, more nuanced inquiries.

## 3.2   1990–2000: Internationalization of Environmental Issues

Since 1990, there has been a notable shift in the focus of environmental diplomacy, with an increasing number of publications examining environmental issues in several countries other than the US and Western Europe. Developing countries like China started adopting sustainable development strategies to balance economic growth with environmental protection, representing a significant shift from their earlier focus on rapid economic development at any cost and aiming to enhance China's global image as a responsible and cooperative actor in international affairs (China's Agenda 21 1994). Meanwhile, hazardous waste management in Brazil has been shown to be complex due to its geographical size, limited expertise, high costs, and technical requirements (Sampaio 1991). Therefore, a regional synoptic approach is necessary while acknowledging the socio-political hurdles that must be overcome through environmental diplomacy and effective communication (Sampaio 1991).

The agenda of environmental cooperation also reached the Mediterranean (Chircop 1992) and Arctic regions (Roginko and LaMourie 1992), prompting UN specialized agencies to promote the idea of sustainability in these regions. However, environmental cooperation faces several obstacles, such as increasing intricacies, challenges surrounding regionalism's boundaries, persistent absence of widespread support for regional agreements, lack of funding, and the difficulties associated with implementing them (Roginko and LaMourie 1992; Chircop 1992).

Despite the emergence of studies focusing on Global South countries, this decade is characterized by disparity in the distribution of research and economic policy analysis on global environmental issues, with most of such work being conducted in the Global North and significant underrepresentation of research efforts originating from the Global South. Moreover, recommendations developed in the Global North and introduced to developing countries could potentially lead to inappropriate policy recommendations and ineffective solutions. Gutman (1994) observed an overestimation of Global South's emissions, which could have profound implications for formulation and negotiations of climate policy. More region-specific and contextualized policy recommendations considering the economic disparities between Northern and Southern countries are required (Gutman 1994).

However, there are challenges in securing international environmental agreements. Firstly, climate science, like the understanding of ozone layer depletion in the past, may involve complexities and uncertainties that can challenge policy formulation and international cooperation (Benedick 1991). Secondly, economic concerns and the interests of industries may run counter to environmental goals, making it challenging to secure international agreements (Benedick 1991). The Montreal Protocol is known for its effectiveness in addressing the ozone layer issue through international cooperation and regulations, but there is a path forward in negotiations toward a Climate Convention (Benedick 1991). Nevertheless, during this period, organizations started incorporating environmental agreements into the trade regimes that posed challenges and opportunities. For the ASEAN, it is an opportunity for countries to leverage their position in global and regional trade to advance their environmental objectives, but at the same time, this may imply that member countries have to relinquish some degree of autonomy in their environmental policies to comply with international standards and agreements (Montes and Magno 1997). Similarly, Lessons from the experience of the Commission for Environmental Cooperation (CEC) in the Americas highlight the need to balance a broader mandate to facilitate funding and autonomy challenges in environmental agreements (Mumme and Duncan 1998).

During this period, the role of the environment in peace and security was increasingly recognized. Brock (1991) identified four key linkages—causal, instrumental, definitional, and normative—that could positively and negatively impact peace and security, including militarization. Brock (1991) suggested that environmental cooperation has historically depended on overall international relations but may gain independent influence over world politics. For example, Molvær (1990) suggested that environmental cooperation among the nations of the Horn of Africa could lead to political and diplomatic interactions, thereby enhancing stability and security. Molvær (1990) further recognized that environmental degradation might not be the sole trigger of conflicts in the Horn of Africa but could be a contributing factor alongside cultural differences and the perception of development as a zero-sum game.

One of the cases that offers important lessons for international cooperation and international environmental policy is the Brazilian Amazon. Deforestation in the Amazon is a politicized issue linked to global inequalities, sovereignty, and funding transfer (Kolk 1998). While the involvement of interactional organizations fostered cooperation, advocates campaigned for the integration of Brazilian sensitivities, playing a pivotal role in compelling institutions like the World Bank to revise their policies to transform conflict into cooperation (Kolk 1998). However, these efforts are often most effective when states are not in acute conflict, targeted at a single country (Kolk 1998). Ecchia and Mariotti (1998) proposed a more active participation of international institutions, including interventions in framing strategic interactions between countries, which can be reached by setting negotiation rules amidst differing priorities and having the ability to influence the actual agreement.

But why do some countries willingly abide by international environmental cooperation while others are reluctant to do so? Is there any theory to explain this? The interest-based theory suggests that countries engage to the extent they are directly or indirectly affected by the environment (Sprinz and Vaahtoranta 1994). Therefore, based on their environmental vulnerabilities, countries can be active, reluctant, passive, or neutral supporters of strict international environmental regulations (Sprinz and Vaahtoranta 1994).

### 3.3 2000–2010: Contextualizing the Shift Towards Multilateral Environmental Diplomacy

Environmental diplomacy evolved from bilateral or regional efforts to broader multilateral engagements during this period. The literature on environmental diplomacy in this period sheds light on the challenges and approaches to addressing environmental issues, particularly climate change, and the differences and similarities between developed and developing countries. Grant and Papadakis (2004) point to differences between Australia and the European Union (EU), especially in terms of the use of market mechanisms and implementation of punitive compliance systems and examining more deeply than simply categorizing as "laggards" or "leaders," there are more similarities than differences. Therefore, cooperation between the regions is required. On the other hand, despite apparent ideational and institutional advances, some regional organizations, such as ASEAN, have been unable to respond effectively to regional environmental challenges due to normative (related to values and principles) as well as material (related to resources and practical constraints) reasons limitations (Elliott 2003). This suggests that this degradation is seen by some as a failure of regional cooperation and a crisis of regional identity and credibility within ASEAN, indicating the region's inability to address environmental challenges has raised questions about its effectiveness and integrity (Elliott 2003). Meanwhile, the US, once a pioneer in domestic environmental lawmaking and a leader in environmental diplomacy, has fallen far behind several European states and the EU that have started implementing robust policies to reduce their greenhouse gas emissions (Harris 2009). The US demonstrates the role of a political administration in transforming the country from an environmental leader to an environmental laggard.

With regard to the issue of global pollution and emissions reduction, diplomacy is crucial in overcoming countries' limited incentives to act due to their small individual impact on emissions, potentially leading to a prisoner's dilemma (Lange and Vogt 2003). Each country prioritizes its interests, hoping others will reduce emissions, resulting in increased global pollution. Therefore, the question arises: Can nonmaterial incentives be applied? Lange and Vogt (2003) suggest that fairness preferences can foster cooperation. Developing countries emphasize equity in addressing global warming, while environmental interest groups in developed nations advocate for equal per capita emissions. These factors can significantly influence national policies and lead to substantial cooperation rates. However, in continuous emission games, fairness alone may not significantly improve cooperation as countries might prioritize increasing absolute payoffs over equity concerns (Lange and Vogt 2003). Countries tend to prefer cooperative regimes that address transboundary pollution while preserving their sovereignty and economic interests (Yoon 2008).

The sovereignty of states makes external enforcement of international environmental agreements challenging, with each nation generally responsible for enforcing its commitments. Given the limitations of external enforcement, international environmental agreements need to be designed in a way that encourages self-enforcement (Wagner 2001). To promote participation and compliance, international environmental agreements often include various mechanisms, such as financial transfers (providing benefits to participating countries), sanctions (penalties for non-compliance), and linkage to other negotiation topics in international politics (linking environmental issues to other important diplomatic matters) (Wagner 2001). This underscores the nuanced nature of international environmental cooperation, where the interplay between equity considerations, national interests, and cooperation mechanisms shapes the strategies and outcomes of participating countries.

Participation in multilateral environmental agreements is also linked to trade openness (Neumayer 2002). Countries are more likely to engage in multilateral environmental cooperation when it benefits their trade relationships (Neumayer 2002). Relatedly, countries deeply involved in international environmental agreements are likely to engage in global trade and asset exchange (Rose and Spiegel 2009). This implies that the decision to join or withdraw from such agreements should consider both direct environmental consequences and indirect economic benefits or costs (Rose and Spiegel 2009). Along with these factors, trust plays a crucial role in international environmental cooperation (Tennberg 2007). Trust is observed as a critical factor in improving domestic environmental policies and enhancing a country's standing in international environmental cooperation efforts. The benefits of such cooperation efforts are enhanced if there is mutual trust between all parties involved (Žičkienė 2007).

During this decade, research articles typically call for a more balanced and equitable approach to addressing climate change, considering the perspectives and needs of developing countries. This raises important questions about the global climate governance system and the role of different nations in mitigating climate change while promoting sustainable development (Najam et al. 2003). Perhaps the most intriguing finding is that future international climate agreements might incorporate a combination of equity principles, including the polluter-pays, egalitarian, and poor losers rules (Lange et al. 2007). This implies that negotiations may need to strike a balance between these principles to gain broader support and achieve meaningful agreements.

Overall, the scholarly literature during this decade underscores the complexities of environmental diplomacy, the role of equity, sovereignty, and trust, and the need for balanced and cooperative approaches to address global environmental challenges.

## 3.4 2010–2023: Advancing Environmental Diplomacy Through Peace and Equality

The intersection of environmental diplomacy with peace and equality became increasingly important during this period and across an array of themes, such as green diplomacy, economic considerations, the relationship between peace and the environment, water diplomacy, and women's representation.

First, a significant number of studies on green diplomacy are defined as the "specialized international development cooperation at a bilateral, regional or global scale" (Ioan 2013; Hammad and Mohammad 2022). Results suggest that green diplomacy strengthens and influences international relations by constructing preventive measures surrounding ecological reservations caused by dangerous or ecologically damaging pollution (Hammad and Mohammad 2022). Green diplomacy represents a new way of diplomacy, which considers not only harmonizing the interests of states but also the shared responsibility of all of humanity—from individuals to communities and states (Ioan 2013).

The second issue concerns the economic impact of environmental diplomacy. For instance, does multilateral environmental diplomacy affect $CO_2$ emissions, considering factors like capital formation, renewable energy consumption, and economic growth? Khan and Hou (2021) explain that environmental degradation initially worsens but then improves as a country's income and development increase. These findings align with the hypothesis that as international commitments and diplomatic relationships strengthen, $CO_2$ emissions start to decline (Khan and Hou 2021). However, although environmental diplomacy initially reduces $CO_2$ emissions in developing countries, in the long run, when signatories do not adhere to treaty requirements, $CO_2$ emissions increase as more treaties are signed (Li et al. 2020). This suggests that signing environmental treaties may be more about international diplomacy than achieving actual results in climate change mitigation (Li et al. 2020). Therefore, countries should focus on fulfilling their treaty obligations rather than engaging in the annual ritual of signing treaties.

Imposing economic sanctions has become a popular tool to influence opponents. But are there any repercussions of economic sanctions on the environment? While sanctions may not directly cause environmental problems, they significantly contribute to environmental degradation by impacting a sanctioned state's economy and trade. Sanctions hinder access to technology and goods, leading to a heavier reliance on natural resources, which harms the environment (Madani 2020). Madani (2020) argues that both the sanctioning and sanctioned parties should be held accountable for these unintended environmental consequences, especially in light of human rights violations caused by sanctions. The lasting and transgenerational environmental impacts of sanctions emphasize the need to consider "environmental justice" in future studies and improve humanitarian exemptions to address environmental concerns when sanctions are used (Madani 2020). Diplomats can play a significant role in these discussions through advocacy, mediation, policy advocacy, monitoring, and the advancement of environmental justice.

Third is the relationship between peace and the environment. Ide (2019) underscores that environmental cooperation can reduce violence within states and foster symbolic reconciliation within and between states, but its effectiveness hinges on contextual factors. Key mechanisms include enhancing trust, understanding, and institutional development (Ide 2019). Despite earlier conceptualization and growing recognition of positive peace, which is understood as preventing violence through structural and social justice reform that addresses the inequitable distribution of power and resources (Galtung 1969), much of the available literature understands peace in negative terms as the mere absence of acute conflict, referred to as negative peace. The prevailing notion of peace, characterized by the absence of violence, has faced criticism within various academic disciplines, including international relations (Diehl 2016), political geography (Williams and McConnell 2014), and peace and conflict studies (Gleditsch et al. 2014). This criticism highlights that concentrating solely on negative peace limits our understanding of the shifts from mere violence avoidance to more constructive types of interaction, such as economic integration or the development of security communities (Adler and Barnett 1998).

However, it is crucial to note the potential adverse effects of environmental peacebuilding. Ide (2019) discusses the "dark side" of environmental peacebuilding, identifying six categories of adverse effects: depoliticization, displacement, discrimination, deterioration into conflict, delegitimization of the state, and degradation of the environment. Addressing these issues is essential to fully realizing the potential of environmental peacebuilding (Ide 2019).

Moreover, Ide (2018) provides insights into environmental peacemaking between states, suggesting that cooperative environmental agreements can positively impact reconciliation between rival states. This effect, however, is contingent on conditions such as high environmental attention, internal political stability, and ongoing reconciliation processes. Ide's findings indicate that environmental challenges also create opportunities for peacemaking and peacebuilding between states (Ide 2018).

The value of positive peace is manifested in water. For the period between 1956 and 2006, a higher number of positive water-related interactions in the previous ten years makes a shift toward more peaceful interstate relations more likely, provided states are not in acute conflict (Ide and Detges 2018). Structural violence, such as dominance by a powerful riparian state and unequal allocation agreements, can hinder transboundary water cooperation. As a result, achieving positive peace in this context involves addressing these inequalities. Conflict prevention in transboundary water cooperation should involve institutionalizing processes and policies that promote positive norms over time, creating systems to prevent conflicts. Furthermore, equity and sustainability should be considered in transboundary water cooperation, as they contribute to positive peace by ensuring fair and sustainable resource distribution. Lastly, it suggests that political will and diplomatic engagement are crucial for fostering positive peace alongside structural changes (Smith and Winterman 2022).

The importance of women's representation in environmental cooperation has gained significant attention in recent years. Scholars argue that environmental issues must be understood as gendered (Alston and Whittenbury 2012; Dankelman 2010; Skinner 2011). The underrepresentation of women in institutions governing global environmental change has faced scrutiny from both scholars and professionals. This underrepresentation contradicts the discourse highlighting "women as vulnerable or virtuous in relation to the environment" regarding climate change (Arora-Jonsson 2011). Consequently, the UNFCCC parties have adopted resolutions in 2002 and 2013 to enhance women's engagement in UNFCCC negotiations.

Several factors influence women's representation in state delegations to international environmental negotiations, such as the UNFCCC. Higher economic development and political gender equality within the country are associated with a greater representation of women in state delegations (Kruse 2014). Regional factors also play a role, with Eastern Europe and Latin America positively linked to women's representation in delegations, while the Middle East is negatively associated (Kruse 2014). Therefore, addressing gender equality and promoting development can enhance women's representation in international climate change negotiations, contributing to more inclusive and effective decision-making processes in addressing climate change challenges.

To sum up, more recent literature related to environmental diplomacy emphasizes the positive impact of green diplomacy but debates the effectiveness of multilateral agreements on $CO_2$ emissions and scrutinizes the unintended environmental consequences of economic sanctions. Several studies discuss the relationship between peace and the environment, including in the context of water-related issues. Efforts to increase women's representation in global environmental governance are also noted. Overall, the literature demonstrates the evolving diplomatic approaches to address environmental challenges.

# 4  Conclusion

Environmental diplomacy is an important and well-researched area of international relations. While a substantial body of literature on environmental diplomacy exists, it suffers from fragmentation across various disciplines and lacks a cohesive framework. Despite numerous publications spanning decades and covering different aspects like environmental cooperation, climate diplomacy, and water diplomacy, there is a need for a more systematic approach to understanding and integrating these diverse perspectives.

This chapter synthesized the literature on environmental diplomacy, tracing its evolution from its narrow conceptualization in the 1980s to its internationalization in the 1990s, adoption in multilateral environmental agreements in the 2000s, and its relationship with issues related to peace and equality since 2010.

Notably, the existing literature tends to refer more to environmental cooperation, which is distinct from environmental diplomacy. Therefore, the literature on environmental diplomacy that examines how it might be conducted and its implications for governance and global environmental policy is relatively sparse. There could be several reasons for this. Firstly, the literature on environmental diplomacy is spread across fields such as political science, environmental science, and international relations. This fragmentation hampers a cohesive understanding and limits comprehensive analysis. Secondly, there is a recognized absence of a structured framework to synthesize the various aspects of environmental diplomacy. Terms like environmental cooperation, climate diplomacy, and others are used interchangeably, reflecting the challenge of synthesizing a cohesive body of knowledge. Furthermore, the lack of research from practitioners directly involved in environmental diplomacy could also add value. More empirical research and case studies on environmental diplomacy involve examining specific instances of successful or unsuccessful environmental negotiations, cooperation efforts, and diplomatic strategies are required. Strengthening scientific research and policy partnerships in the field of environmental diplomacy will help better understand this form of diplomacy, improve its environmental effectiveness, and find new ways to solve global environmental problems.

# References

Adler E, Barnett M (eds) (1998) Security communities (Cambridge studies in international relations, series no. 62). Cambridge University Press

Arora-Jonsson S (2011) Virtue and vulnerability: discourses on women, gender and climate change. Glob Environ Change 21(2):744–751. https://doi.org/10.1016/j.gloenvcha.2011.01.005

Alston M, Whittenbury K (eds) (2012) Research, action, and policy: addressing the gendered impacts of climate change. Springer Science & Business Media. Retrieved from https://books.google.com/books?hl=en&lr=&id=5n93NDe3zDkC&oi=fnd&pg=PR5&dq=info:7dk21kTwiJsJ:scholar.google.com&ots=4RcPAv5TgA&sig=enpJ3bP4xiNAJg3E_eccCCx0dUI

Benedick RE (1991) The diplomacy of climate change: lessons from the montreal ozone protocol. Energy Policy 19(2):94–97. https://doi.org/10.1016/0301-4215(91)90124-7

Betsill MM, Corell E (2008) NGO diplomacy: the influence of nongovernmental organizations in international environmental negotiations. MIT Press, Cambridge

Bilder RB (1972) Controlling great lakes pollution: a study in United States-Canadian environmental cooperation. Mich Law Rev 70(3):469–556. https://doi.org/10.2307/1287537

Biermann F (2007) 'Earth system governance' as a crosscutting theme of global change research. Glob Environ Change 17(3–4):326–337. https://doi.org/10.1016/j.gloenvcha.2006.11.010

Brock L (1991) Peace through parks: the environment on the peace research agenda. J Peace Res 28(4):407–423. https://doi.org/10.1177/0022343391028004006

Carroll JE (1983) Environmental diplomacy: an examination and a prospective of Canadian-U.S. Transboundary environmental relations. University of Michigan Press, Ann Arbor

Carroll JE (1986) Water resources management as an issue in environmental diplomacy. Nat Resour J 26(2):207–220. https://digitalrepository.unm.edu/nrj/vol26/iss2/3/

Chasek PS (2012) Global environmental politics. Westview Press, Boulder

Chircop AE (1992) The Mediterranean sea and the quest for sustainable development. Ocean Dev Int Law 23(1):17–30. https://doi.org/10.1080/00908329209545972

China's Agenda 21 (1994) White paper on China's population, environment and development in the 21st century. China Population Today 11(4):5–8

Dankelman I (ed) (2010) Gender and climate change: an introduction. Routledge, London. https://doi.org/10.4324/9781849775274

Diehl PF (2016) Exploring peace: looking beyond war and negative peace. Int Stud Quart 60(1):1–10. https://doi.org/10.1093/isq/sqw005

Ecchia G, Mariotti M (1998) Coalition formation in international environmental agreements and the role of institutions. Eur Econ Rev 42(3):573–582. https://doi.org/10.1016/S0014-2921(97)00117-7

Elliott L (2003) ASEAN and environmental cooperation: norms, interests and identity. Pac Rev 16(1):29–52. https://doi.org/10.1080/0951274032000043235

Galtung J (1969) Violence, peace, and peace research. J Peace Res 6(3):167–191. https://doi.org/10.1177/002234336900600301

Gutman P (1994) Developing countries and international environmental negotiations: the risks of poorly informed choices. Soc Nat Resour 7(4):389–397. https://doi.org/10.1080/08941929409380874

Grant R, Papadakis E (2004) Challenges for global environmental diplomacy in Australia and the European Union. Aust J Int Aff 58(2):279–292. https://doi.org/10.1080/1035771042000222100123

Gleditsch NP, Vestby J, Strand H (2014) Peace research—just the study of war? J Peace Res 51(2):145–158

Harris PG (2009) Beyond bush: environmental politics and prospects for US climate policy. Energy Policy 37(3):966–971. https://doi.org/10.1016/j.enpol.2008.10.042

Hammad HH, Mohammad AA (2022) The emergence of green diplomacy in international relations: a qualitative study. Croat Int Relat Rev 28(90):311–331. https://doi.org/10.2478/CIRR-2022-003

Ide T (2018) Does environmental peacemaking between states work? Insights on cooperative environmental agreements and reconciliation in international rivalries. J Peace Res 55(3):351–365. https://doi.org/10.1177/0022343317750216

Ide T (2019) The impact of environmental cooperation on peacemaking: definitions, mechanisms, and empirical evidence. Int Stud Rev 21(3):327–346. https://doi.org/10.1093/isr/viy014

Ide T (2020) The dark side of environmental peacebuilding. World Dev 127:104777. https://doi.org/10.1016/j.worlddev.2019.104777

Ide T, Detges A (2018) International water cooperation and environmental peacemaking. Glob Environ Polit 18(4):63–84. https://doi.org/10.1162/glep_a_00478

Ioan S (2013) Green diplomacy—the chance to mitigate the effects of the economic crisis in the context of sustainable development. Procedia Soc Behav Sci 81:224–228. https://doi.org/10.1016/j.sbspro.2013.06.417

Khan I, Hou F (2021) Does multilateral environmental diplomacy improve environmental quality? The case of the United States. Environ Sci Pollut Res 28(18):23310–23322. https://doi.org/10.1007/s11356-020-12005-2

Kruse J (2014) Women's representation in the UN climate change negotiations: a quantitative analysis of state delegations, 1995–2011. Int Environ Agreem 14(4):349–370. https://doi.org/10.1007/s10784-014-9245-6

Kolk A (1998) From conflict to cooperation: International policies to protect the Brazilian Amazon. World Dev 26(8):1481–1493. https://doi.org/10.1016/S0305-750X(98)00062-X

Lange A, Vogt C (2003) Cooperation in international environmental negotiations due to a preference for equity. J Public Econ 87(9):2049–2067. https://doi.org/10.1016/S0047-2727(02)00044-0

Lange A, Vogt C, Ziegler A (2007) On the importance of equity in international climate policy: an empirical analysis. Energy Econ 29(3):545–562. https://doi.org/10.1016/j.eneco.2006.09.002

Li G, Zakari A, Tawiah V (2020) Does environmental diplomacy reduce $CO_2$ emissions? A panel group means analysis. Sci Total Environ 722:137790. https://doi.org/10.1016/j.scitotenv.2020.137790

Mcintire D (2014) Eco-diplomacy: building the foundation. Foreign Service J. Retrieved from https://afsa.org/eco-diplomacy-building-foundation

McConnell F, Megoran N, Williams P (eds) (2014) Geographies of peace: new approaches to boundaries, diplomacy and conflict resolution. I.B. Tauris

Madani K (2020) How international economic sanctions harm the environment. Earths Future 8(12):e2020EF001829. https://doi.org/10.1029/2020EF001829

Molvær RK (1990) Environmental cooperation in the horn of Africa. A UNEP perspective. Bull Peace Propos 21(2):135–142. https://doi.org/10.1177/096701069002100203

Montes MF, Magno FA (1997) Trade and environmental diplomacy: strategic options for ASEAN. Pac Aff 70(3):351–372. https://doi.org/10.2307/2761027

Mumme S (1984) The Cananea copper controversy: lessons for environmental diplomacy. Interam Econ Aff 38(Summer):3–22

Mumme S, Duncan P (1998) The commission for environmental cooperation and environmental management in the Americas. J Interam Stud World Aff 36:41. https://doi.org/10.2307/166423

Najam A, Huq S, Sokona Y (2003) Climate negotiations beyond Kyoto: developing countries concerns and interests. Clim Policy 3(3):221–231. https://doi.org/10.1016/S1469-3062(03)00057-3

Neumayer E (2002) Does trade openness promote multilateral environmental cooperation? World Econ 25:815–832. https://doi.org/10.1111/1467-9701.00464

Roginko AY, LaMourie MJ (1992) Emerging marine environmental protection strategies for the Arctic. Mar Policy 16(4):259–276. https://doi.org/10.1016/0308-597X(92)90044-P

Rose A, Spiegel M (2009) Noneconomic engagement and international exchange: the case of environmental treaties. J Money Credit Bank 41:337–363. https://doi.org/10.1111/j.1538-4616.2009.00208.x

Sampaio JAB (1991) Hazardous wastes management in Brazil: the need for a regional synoptic approach. Water Sci Technol 24(12):11–18. https://doi.org/10.2166/wst.1991.0364

Sinclair M (1986) The environmental cooperation agreement between Mexico and the United States: a response to the pollution problems of the borderlands. Cornell Int Law J 19(1):87–142. https://scholarship.law.cornell.edu/cilj/vol19/iss1/4

Skinner E (2011) Gender and climate change: overview report', bridge cutting edge pack. BRIDGE/Institute of Development Studies (IDS), Brighton

Sprinz D, Vaahtoranta T (1994) The interest-based explanation of international environmental policy. Int Organ 48(1):77–105. https://doi.org/10.4324/9781315092546-9

Smith D, Winterman K (2022) Models and mandates in transboundary waters: institutional mechanisms in water diplomacy. Water 14(17). https://doi.org/10.3390/w14172662

Tennberg M (2007) Trust in international environmental cooperation in Northwestern Russia. Coop Confl 42(3):321–335. https://doi.org/10.1177/0010836707079935

Thompson F (1984) Environmental diplomacy: an examination and a prospective of Canadian-U.S. Transboundary environmental relations, by John E. Carroll. Polit Sci Q 99(1):130–131. https://doi.org/10.2307/2150288

UN (2015) Sustainable development goals. United Nations. Retrieved from https://sdgs.un.org/goals

UNFCCC (2015) The Paris agreement. United Nations framework convention on climate change. Retrieved from https://unfccc.int/process-and-meetings/the-paris-agreement/the-paris-agreement

Wagner U (2001) The design of stable international environmental agreements: economic theory and political economy. J Econ Surv 15:377–411. https://doi.org/10.1111/1467-6419.00143

Wapner P (1995) Politics beyond the state: environmental activism and world civic politics. World Polit 47(3):311–340. https://doi.org/10.1017/S0043887100016415

Yoon E (2008) Cooperation for transboundary pollution in Northeast Asia: non-binding agreements and regional countries' policy interests. Pac Focus 22:77–112. https://doi.org/10.1111/j.1976-5118.2007.tb00298.x

Žičkienė S (2007) Cooperation in environmental governance—a new tool for environment protection progress. Eng Econ: 42–50. https://inzeko.ktu.lt/index.php/EE/article/view/12222